The Urban Book Series

Aims and Scope

The Urban Book Series is a resource for urban studies and geography research worldwide. It provides a unique and innovative resource for the latest developments in the field, nurturing a comprehensive and encompassing publication venue for urban studies, urban geography, planning and regional development.

The series publishes peer-reviewed volumes related to urbanization, sustainability, urban environments, sustainable urbanism, governance, globalization, urban and sustainable development, spatial and area studies, urban management, urban infrastructure, urban dynamics, green cities and urban landscapes. It also invites research which documents urbanization processes and urban dynamics on a national, regional and local level, welcoming case studies, as well as comparative and applied research.

The series will appeal to urbanists, geographers, planners, engineers, architects, policy makers, and to all of those interested in a wide-ranging overview of contemporary urban studies and innovations in the field. It accepts monographs, edited volumes and textbooks.

More information about this series at http://www.springer.com/series/14773

Eduardo Medeiros
Editor

Territorial Cohesion

The Urban Dimension

Springer

Editor
Eduardo Medeiros
DINÂMIA'CET-IUL
Instituto Universitário de Lisboa
(ISCTE-IUL)
Lisbon, Portugal

ISSN 2365-757X ISSN 2365-7588 (electronic)
The Urban Book Series
ISBN 978-3-030-03385-9 ISBN 978-3-030-03386-6 (eBook)
https://doi.org/10.1007/978-3-030-03386-6

Library of Congress Control Number: 2018960210

This Springer imprint is published by the registered company Springer Nature Switzerland AG
The registered company address is: Gewerbestrasse 11, 6330 Cham, Switzerland

Foreword

As laid down in the EU Treaty since 2009, EU policies should promote Territorial Cohesion which, in turn, require a polycentric and balanced territorial development approach. As such, it goes without saying that this book covers a fundamental aspect of implementing Territorial Cohesion goal: the urban dimension and all its related policy elements.

Moreover, the book content instils a needed scientific debate on the Territorial Cohesion concept, which is still largely unexplored in available literature, and especially the importance of several urban-related components (polycentrism, integrated territorial policy approaches and functional regions) to achieving the ultimate EU policy goal of Territorial Cohesion. In this regard, this book contains valuable insights, both from a theoretical and empirical standpoint, on the importance of implementing sound and balanced urban policies to achieving Territorial Cohesion goals.

ESPON Programme has, since 2002, supported the work of an extended network of European researchers and experts to obtain new and substantial evidence on Europe's territorial trends, structures, perspectives and policy impacts. Since the start of ESPON programme, it has supported the elaboration of several studies aiming at identifying the main dimensions, components and indicators of Territorial Cohesion (INTERCO, KITKASP, SeGI, etc.). I am glad that the results of ESPON research continue to be relevant and useful for supporting the policy debate and development of publications on territorial cohesion, such as this book.

In sum, the book content can be of great value for the academia and policy-makers, as it further nurtures the debate around different interpretations of the Territorial Cohesion concept. I deeply recommend reading this book, which offers a novel and refreshed overview of the Territorial Cohesion concept, by identifying and debating urban-related elements, which are crucial to implementing Territorial Cohesion Policies.

Luxembourg

Ilona Raugze
Director at ESPON EGTC

Contents

Editors and Contributors

About the Editor

Eduardo Medeiros is a Geography Professor and a Senior Research Integrated Fellow in DINÂMIA'CET-IUL, Instituto Universitário de Lisboa (ISCTE-IUL), Portugal. He has a Ph.D. in Geography—Regional and Urban Planning, and has more than 30 published papers in international journals, eight books and nine book chapters. His research interests are focused on Territorial Cooperation, Territorial Cohesion, Territorial Development and Territorial Impact Assessment. He is a DG REGIO (European Commission) expert, an URBACT expert, and a Horizon 2020 evaluator. He is also a Regional Studies Association fellow and belongs to its Cohesion Policy Research Network. He has coordinated several international policy evaluation projects and is a member of an ESPON project. He was invited as a project adviser and to write reports by DG REGIO. He was already invited to be a keynote speaker by several International Universities and EU institutions (European Commission and Committee of the Regions). He is part of the scientific and editorial committee of some journals and a peer reviewer of more than 20 international journals.

Contributors

Bjørn Terje Andersen Hedmark County Council, Hamar, Norway

Kai Böhme Spatial Foresight, Heisdorf, Luxembourg

Giancarlo Cotella Interuniversity Department of Regional and Urban Studies and Planning, Politecnico di Torino, Turin, Italy

Martin Ferry European Policies Research Centre, University of Strathclyde, Glasgow, Scotland

Erik Hagen Hedmark County Council, Hamar, Norway

Tomasz Komornicki Department of Spatial Organization, Institute of Geography and Spatial Organization Polish Academy of Sciences, Warsaw, Poland

Christian Lüer Spatial Foresight, Heisdorf, Luxembourg

Jiří Malý Czech Academy of Sciences, Institute of Geonics, Brno, Czech Republic

Eduardo Medeiros DINÂMIA-CET IUL - Instituto Universitário de Lisboa (ISCTE-IUL), Lisbon, Portugal

Paulo Neto Department of Economics, University of Évora, Public Policy Monitoring Unit (UMPP), CICS.NOVA.UÉvora and CEFAGE-UÉ, Évora, Portugal

Oto Potluka Centre for Philanthropy, University of Basel, Basel, Switzerland

Bjørnar Sæther Department of Sociology and Human Geography, University of Oslo, Oslo, Norway

Anabela Santos International Centre for Innovation, Technology and Education Studies (iCite), Solvay Brussels School of Economics and Management, Université Libre de Bruxelles, Brussels, Belgium

Maria Manuel Serrano Department of Sociology, University of Évora, Public Policy Monitoring Unit (UMPP), CICS.NOVA.UÉvora and SOCIUS-CSG/ ISEG-UL, Évora, Portugal

Martin Špaček CETIP Network, Prague, Czech Republic

Maria Toptsidou Spatial Foresight, Heisdorf, Luxembourg

Arno van der Zwet University of the West of Scotland, Paisley, Scotland

Jacek Zaucha Faculty of Economics, University of Gdańsk and Institute for Development, Gdańsk and Sopot, Poland

Introduction

What is Territorial Cohesion? This question has been around both in academic and political arenas and meanders, since this European concept was introduced into the EU Policy Agenda, around the turn of the millennium, as a French policy rationale (Aménagement du territoire). The answer to this question varies from author to author. But one key aspect is consensual: Territorial Cohesion implies the need to achieve a more balanced and harmonious territory, as implied in the EU Treaties, since the Treaty of Rome (1957). However, it was only by 2009, with the entry into force of the Lisbon Treaty, that the former article 158°, now article 174°, clearly expresses that, 'in order to promote its overall harmonious development, the Union shall develop and pursue its actions leading to the strengthening of its economic, social and territorial cohesion'.

Indeed, whilst the EU policies main aims of promoting the twin community objective of economic and social cohesion have been around since the mid-1980s (Single European Act), the recognition of perennial and widening (due to successive accession of new Member States) territorial disparities has led to the inevitable inclusion of the Territorial Cohesion as an EU mainstream policy goal, alongside with the goals of promoting socio-economic cohesion. One can question, however, if this is not redundant policy rationale. In our view, it is, since the achievement of Territorial Cohesion, generically understood as a goal to achieving a more balanced and harmonious territory, requires both positive economic and social cohesion trends within a given territory. We also agree with the idea expressed in the Second EU Cohesion Report, which claimed that Territorial Cohesion goes beyond the notions of Social and Economic cohesion. For the most part, however, there is a widespread tendency to only add 'environmental policy concerns' to the former two analytic dimensions (social and economic) when debating the concept of Territorial Cohesion, to complement its conceptual rationale.

Territorial Cohesion remains an elusive concept, reflecting emerging axioms that have begun to permeate academic discourse in the past years. There is an increasing recognition that this concept should include other fundamental analytic dimensions, such as the Territorial Governance/Cooperation and the Polycentrism. The latter was invoked by the European Spatial Development Perspective (ESDP—1999), as a vital goal of an EU spatial development policy: 'development of a balanced and polycentric urban system and a new urban-rural relationship'. Since then, however, there has been a lack of a rich vein of theoretical reasoning devoted to debate the importance of this (urban polycentric system) and other aspects associated with the 'urban dimension' of Territorial Cohesion policies. In this context, this book intends to fill this literature void, by shedding light on some fundamental urban policy components, which can contribute to achieving such policies.

In more detail, this book has nine chapters and is organised into three separate parts, each composed of three chapters. In the first part—'Territorial Cohesion and the Urban Dimension'—Eduardo Medeiros (the Book Editor)—which has dedicated a significant part of his academic research (since 2003) to the analysis of the Territorial Cohesion concept—identifies and debates some crucial policy aspects associated with the urban dimension of Territorial Cohesion. In the following chapter, Oto Potluka and Martin Špaček highlight the importance of engaging the urban civil society as a means of promoting territorial integrated approaches to regional and urban development, to achieve Territorial Cohesion Policies. Subsequently, in the third chapter, Jacek Zaucha and Tomasz Komornicki examine how cities contribute to the urban dimension of territorial cohesion (polycentricity, territorial governance and the socio-economic growth), whilst debating the role of cities and Functional Urban Areas, as producers of well-being and prosperity in space.

The second part of the book—'Territorial Cohesion and Urban Balanced Systems?'—starts with a chapter in which Jiří Malý sheds deeper light on the importance of the implementation of polycentric urban development, as a normative and ideological vision for a more balanced spatial development. In a complementary way, Bjørnar Sæther is responsible for Chap. 5 where, together with Erik Hagen and Bjørn Terje Andersen, he debates the importance of cross-border functional regions to archiving Territorial Cohesion Processes, in cross-border and transnational regions. Finally, this part of the book ends with a chapter written by Arno van der Zwet and Martin Ferry, dedicated to better understand the added value of Integrated Sustainable Urban Development Strategies to achieving Territorial Cohesion policy goals.

Finally, the third and last part of the book—'Territorial Cohesion and Urban Development Policies'—starts with a chapter written by Giancarlo Cotella, which is mostly focused on the relation between the implementation of EU Cohesion Policies and Urban Agendas at the national scale, and their importance to follow a more cohesive spatial and urban planning path. Conversely, in the following chapter, Paulo Neto, Maria Manuel Serrano and Anabela Santos, set their attention on analysing how policies directed to cities and urban matters, and more specifically the Urban Agenda for the EU, have become more relevant over the past decades, in the European Union's public policy set. This book part ends with a final chapter, written by Kai Böhme, Christian Lüer and Maria Toptsidou, which alert to current urban growth trends in the European territory, which do not necessarily favour Territorial Cohesion policy goals.

Eduardo Medeiros

Part I
Territorial Cohesion and the Urban Dimension

Chapter 1
Debating the Urban Dimension of Territorial Cohesion

Eduardo Medeiros

Abstract The Territorial Cohesion goal was only included in the EU Treaty by 2009, with a view to promote a more balanced and harmonious European territory. One year earlier (2008), the European Commission (EC) published the 'Green Paper on Territorial Cohesion—Turning territorial diversity into strength'. Neither one, nor the other, clearly defines the meaning of the Territorial Cohesion concept. The later, however, proposes three main policy responses towards more balanced and harmonious development: (i) Concentration: overcoming differences in density; (ii) Connecting territories: overcoming distance; and (iii) Cooperation: overcoming division. Although not explicitly, this document identifies several 'urban questions' to be dealt when promoting territorial cohesive policies: avoiding diseconomies of very large agglomerations and urban sprawl processes, combating urban decay and social exclusion, avoiding excessive concentrations of growth, promoting access to integrated transport systems and creating metropolitan bodies. In this light, this chapter proposes to debate the importance of the urban dimension to achieve the goal of territorial cohesion at several territorial levels.

Keywords Territorial cohesion · Urban dimension · Polycentrism
Functional regions · EU cohesion policy

1.1 The Urban Dimension: A Key Pillar to Achieve Territorial Cohesion

With the entering into force of the Lisbon Treaty (2009), Territorial Cohesion became a reinforced political objective of the European Union (EU). In sum, this Territorial Cohesion rationale supports the longstanding EU political goal of achieving an overall harmonious and balanced development of the EU territory, by reducing disparities between its various regions and the backwardness of the least favoured ones, from a

E. Medeiros (✉)
DINÂMIA-CET IUL - Instituto Universitário de Lisboa (ISCTE-IUL), Lisbon, Portugal
e-mail: eduardo.medeiros@iscte-iul.pt

E. Medeiros (ed.), *Territorial Cohesion*, The Urban Book Series,
https://doi.org/10.1007/978-3-030-03386-6_1

socio-economic standpoint (article 174 of the Treaty on the Functioning of the EU) (Dao et al. 2017).

As a concept, Territorial Cohesion is relatively new, and is still mostly an EU concept (Medeiros 2016a), echoing its French roots (Faludi 2004) and its formal introduction in the Amsterdam Treaty (1999) and, more recently, in the Lisbon Treaty (2009) (Servillo 2010). Also noteworthy, Territorial Cohesion is generally understood as a contested, illusive and vague concept (Colomb and Santinha 2014; Davoudi 2005; Faludi 2005; Zillmer et al. 2012). Indeed, being an umbrella concept (Faludi 2007) makes its meaning and fundamental analytic dimensions vary from author to author (Holder and Layard 2011; Medeiros 2014; Schön 2005; Van Well 2012). Fundamentally, this has been the case since the first published dissertations on Territorial Cohesion were presented, one being a study from the Committee of the Regions (COR 2003), and another a full dedicated section on the Second EU Cohesion Report (EC 2001). Here, while the former focused on the economic and social elements of cohesion, the later associated, for the first time, the notion of Territorial Cohesion with the EU goal of achieving a more balanced and harmonious development, following the rationale presented in the formulation of the European Spatial Development Perspective (ESDP) (EC 1999).

Crucially, the Second Cohesion Report introduced the 'urban element' in the Territorial Cohesion debate, namely, by invoking the advantages of achieving a more polycentric development (Faludi 2006), as a basis for a more balanced distribution of economic activities across the EU, as expressed in the ESDP. Since then, the EU Territorial Cohesion Policy rationale has emerged as a policy paradigm of the 'EU territorial development' (Clifton et al. 2016). At the same time, one started to see the emergence of distinct complementary elements to the common socio-economic ones when invoking the notion of Territorial Cohesion, such as environmental aspects (EEA 2010), soft spaces of spatial development (Luukkonen and Moilanen 2012), the valorisation of territorial capital (Vanolo 2010) and, ultimately, the need for a more spatially balanced and sustainable development process (González et al. 2015).

In a nutshell, many of the proposed definitions for the Territorial Cohesion concept encompass elements associated with economic, social and environmental aspects of development policies (see Medeiros 2016a). For instance, the components for understanding Territorial Cohesion in the Tequila Model are mostly associated with the 'economy-society-environment dimensional triad' (ESPON 3.2, 2006; Bradley and Zaucha 2017). Even so, a few (integrated and balanced territorial systems, compact city form—reduction of urban sprawl, efficient and polycentric urban systems, etc.) are clearly associated with an urban dimension of Territorial Cohesion. In a relatively older model, known as the 'star of territorial cohesion', the author proposed the 'polycentrism' and 'territorial cooperation' elements (later on associated with the notion of territorial governance) as key dimensions for the understanding of the Territorial Cohesion concept, alongside the socio-economic and environmental sustainability dimensions (Medeiros 2005) (Fig. 1.1).

Based on this star model, we understand the concept of territorial cohesion as 'the process of promoting a more cohesive and balanced territory, by: (i) supporting the reduction of socioeconomic territorial imbalances; (ii) promoting environmental

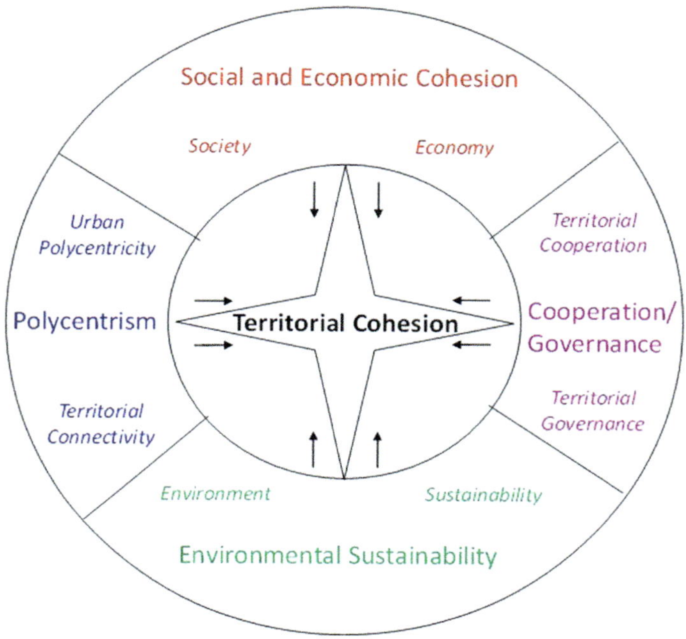

Fig. 1.1 The star of Territorial Cohesion. *Source* Medeiros (2016a)

sustainability; (iii) reinforcing and improving the territorial cooperation/governance processes; and (iv) reinforcing and establishing a more polycentric urban system' (Medeiros 2016a: 10). More acutely, for the morphologic polycentricity dimension, we proposed four main components, mostly associated with the urban systems characteristics in a given territory: (i) hierarchy/ranking, (ii) density, (iii) connectivity and (iv) distribution/shape.

Likewise, elements associated with this urban element of territorial cohesion are also debated in the Green Paper on Territorial Cohesion, namely, associated with the need to avoid excessive concentration (overcoming differences in density), and the need to correct urban sprawl and urban decay trends (EC 2008). Rather expectedly, the ESPON (European Spatial Planning Observatory Network) programme, established in 2002, supported the elaboration of several reports which discussed the main dimensions, components and respective indicators of the Territorial Cohesion concept (Abrahams 2014). The first, gave way to the elaboration of the already mentioned TEQUILA model (ESPON 3.2, 2006). Soon after, the INTERCO (2011) and the KITCASP (2012) reports proposed concrete indicators to assessing Territorial Cohesion trends. The former, identified one of its main Territorial Cohesion dimensions as the 'integrated polycentric territorial development', whilst associating it with territorial cooperation, migration, density and polycentricity-related indicators. Instead, the KITCASP project proposed the 'Integrated Spatial Development'

Table 1.1 The urban dimension of territorial cohesion in available literature

Model/Report	Dimension	Components
The Territorial Cohesion Star	Polycentrism	– Hierarchy/Ranking – Density – Connectivity – Distribution/Shape
ESPON Tequila Model		– Compact city form; Reduction of sprawl – Efficient and polycentric urban system – Cooperation between city and countryside – Integrated and balanced territorial system
ESPON KITCASP	– Promote polycentric and balanced territorial development – Encouraging integrated development in cities, rural and specific regions	– Improve settlements' performance in European and global competition and promote economic prosperity towards sustainable development – Contribute to reducing the strong territorial polarisation of economic performance, avoiding large regional disparities by addressing bottlenecks to growth – Smart development of city regions at varying scales – Development of the wide variety of rural areas to take account of their unique characteristics; and – Recognise and promote urban–rural interdependence through integrated governance and planning based on broad partnership
ESPON INTERCO	Integrated polycentric territorial development	– Population potential within 50 km – Net migration rate – Cooperation intensity – Cooperation degree – Polycentricity index

Source Several—own elaboration

main dimension of Territorial Cohesion to complement the economic (Economic Competitiveness and Resilience), social (Social Cohesion and Quality of Life) and environmental (Environmental Resource Management) related dimensions. When it comes to indicators, however, it proposes the use of population density/change, modal split, land-use change, house completions and access to services, which, in our opinion, are far less appropriate to asses this dimension than the ones advanced by the INTERCO project (Table 1.1)

Clearly, the Territorial Cohesion concept builds upon the notions of economic and social cohesion (Janin Rivolin 2005) whilst its urban dimension is still not sufficiently understood and clarified. As it turns out, however, the rising concerns on the implementation of urban development and regeneration policies in EU Member States (Colomb and Santinha 2014) could lead to an increasing need to highlight the importance of the urban dimension to achieve the goals of Territorial Cohesion Policies. For the most part, the urban dimension of Territorial Cohesion can be associated with the benefits related with a more balanced and harmonious distribution of the urban settlement across a given territory. This idea has strong connotations with the notion of 'Polycentricity' 'that encourages regions and cities, working with neighbouring territories, to explore common strengths and reveal potential complementarities, which brings added value that cannot be achieved by the individual regions and cities in isolation' (ESPON 2016: 1).

In general terms, the role of urban areas in promoting Territorial Cohesion processes is strong and diverse, mainly in highly urbanised territories, as is the case of the EU territory. Strong because cites are innovation hubs and places where global challenges can be best tackled. For instance, they have the potential to produce renewable and clean energy and to reduce energy consumption and CO^2 emissions, 'as the density of urban areas allows for more energy efficient forms of housing and transport' (EC 2014b: 4). As a recent ESPON report asserts 'cities play an important role in regional development and as part of strategies to promote territorial cohesion. Cities are functional areas where population and economic activities are concentrated and around which flows of goods and persons are organised. They are nodes in urban networks at different scales from the regional to the global. In both these respects, cities and towns of all sizes play important roles as drivers of territorial development' (ESPON 2014a: 7). As the term implies, however, the urban future requires developing innovative, social inclusive, eco-friendly and intelligent intra-urban and inter-urban transport and energy systems, and wider civic participation (Nijkamp and Kourtit 2013). On the whole, the urban dimension of Territorial Cohesion encompasses at least eight main components, with evident inter-relationship (Table 1.2).

Unlike other parts of the world (read USA and China), Europe is characterised by less concentrated and more polycentric urban structures (Fig. 1.2), and by dense networks of small and medium towns (EC 2014b). There is, nevertheless, a more densely populated axis known as the 'Pentagon area', consisting of the urban areas of London, Paris, Milan, Berlin and Hamburg, characterised by higher levels of socio-economic development. Hence, at the EU scale, urban development policies should foster the territorial competitiveness of the EU territory also outside this area, as the 'polycentric and balanced territorial development of the EU is a key element of achieving territorial cohesion' (Territorial Agenda 2011: 7).

To some degree, making Europe more polycentric has the potential to 'unleashing regional diversity and endogenous development as well as territorial cooperation as means to optimise the location of investments and reduce regional disparities, to support balanced and polycentric urban structures, favouring compact settlements and smart renewal of cities, as well as a sustainable management of natural and

Table 1.2 Main Components of the Urban Dimension of Territorial Cohesion

Component	Main goal	Importance for Territorial Cohesion
1—Polycentric and balanced urban development	– Favour a more balance distribution of the population across the territories	– Increase the ability to explore the territorial potentials across the territories. – Mitigate the negative effects of excessive concentration of inhabitants (pollution, criminality, traffic jams, stress, etc.)
2—Functional Urban Areas	– Favour territorial complementarities	– Increase territorial efficiency
3—Integrated Urban Development	– Favour territorial efficiency and sustainability	– Increase environmental sustainability – Increase territorial efficiency – Improve territorial governance processes
4—Social Cohesive Urban Development	– Favour socio-economic cohesion	– Reduce poverty – Reduce social imbalances
5—Urban connectivity	– Favour territorial accessibility	– Increase urban mobility – Increase environmental sustainability
6—Urban morphology – compact vs urban sprawl	– Favour urban compactness vs urban sprawl	– Increase territorial efficiency – Increase territorial connectivity – Increase territorial sustainability
7—Urban resilience	– Favour urban capacity to face adversities	– Increase territorial resilience – Increase territorial sustainability
8—Urban Planning	– Favour the anticipation of future scenarios	– Increase territorial modernisation – Increase territorial organisation

Source Own elaboration

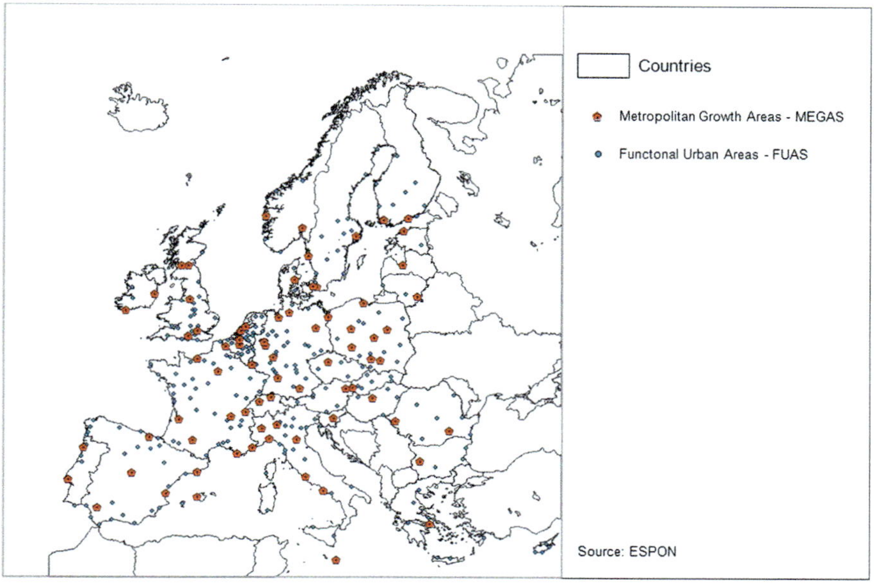

Fig. 1.2 The European Urban System—own elaboration based on ESPON data

cultural resources' (ESPON 2014b: 12). This policy goal is achieved, at the EU level by reinforcing second-tier cities, and a more developed network between cities at all territorial levels, together with a growing interdependency of urban and rural areas.

The potential for a more resource efficient habitat has long been recognised. Nevertheless, the policy response at the national and European levels to integrate sectoral initiatives has been slow, at best (EC 2014b). As Nijkamp and Kourtit postulate (2013) 'cities increasingly act in a system of connected networks that serve as strategic alliances for the development of our world. In this perspective, urban agglomerations are not necessarily a source of problems, but offer the integrative geographic action platform for creative solutions and new opportunities'. Indeed, the realisation among national governments and international agencies of the importance of urban centres to sustainable development has rapidly gained currency in recent years (Satterthwaite 2016).

Similarly, the notion of functional cities relates to the concept of Territorial Cohesion as it facilitates the identification of adequate urban influential areas with potential functional complementarities, which is required to better define sound Territorial Cohesion policies. This concept resonates with the term of Functional Urban Areas (FUA), which 'can be described by its labour market basin and by the mobility pattern of commuters, and includes the wider urban system of nearby towns and villages that are economically and socially dependent on an urban centre' (URBACT 2014: 13). These FUAs can also extrapolate national borders, thus making a case to function as cross-border FUAs, whilst contributing to reinforcing territorial governance processes across administrative boundaries. As the ESDP points out 'promoting complementarity between cities and regions means simultaneously building on the advantages and overcoming the disadvantages of economic competition between them' (EC 1999: 21). As seen in both Figs. 1.2 and 1.3, the identification of FUAs varies from source to source, but they entail common criteria of being associated with large urban agglomerations.

At the urban level, there are also concrete measures which can lead to an increase in levels of Territorial Cohesion. These include, for instance, the correction of urban sprawl processes, as it contributes to high and unsustainable energy consumption rates (Territorial Agenda 2011). This goal is closely related with the need to implement sound and strategic urban planning, linked with wider regional planning strategies, as proposed by the UE Urban Agenda (2016). Additionally, urban development policies should favour the implementation of socio-economic cohesion, environmental sustainability, sound governance and urban resilience measures, in an interconnected way.

In combination with the urban elements already debated, the contribution of Integrated Sustainable Urban Strategies for promoting Territorial Cohesion is also paramount. For one, this policy rationale represents the potential to improve environmental sustainability development paths. On the other hand, and according to the EU Integrated Territorial Investment rationale, it promotes sounder territorial governance processes, as it allows for local authorities to initiate wider partnerships with other local authorities (Glinka 2017). For instance, the Polish experience so

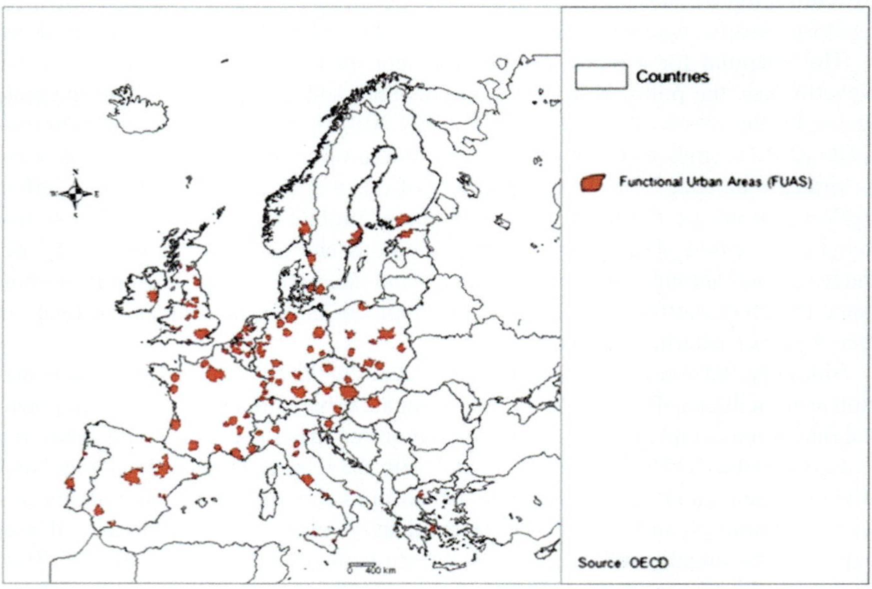

Fig. 1.3 Functional Urban Areas in Europe—own elaboration based on OECD data

far on the implementation of Integrated Territorial Investments has been reserved only for regional capitals and their functional areas. As such, a new mechanism (Integrated Territorial Agreement) was designed to be implemented in medium-sized towns and their functional areas, in order to better exploit their development potential and to mitigate their development bottlenecks (Kamrowska-Zaluska and Obracht-Prondzynska 2017).

1.2 EU Policies and the Urban Question

By today, approximately 72% of the EU population dwell in urban areas (cities, towns and suburbs). In spite of this, the EU does not have any formal Council formation dedicated to urban policies. Even so, 'there is an explicit agreement at European level on the character of the European city of the future and the principles on which an ideal European city should be based' (EC 2014b: 6). In practice, the urban question has been addressed by EU Cohesion Policy, directly or indirectly, namely, through the launching of the URBAN Community Initiative (CI) in 1994 (Chorianopoulos and Iossifides 2006), the national Operational Programmes and, more recently, the Integrated Sustainable Urban Development Strategies. In almost every way, the URBAN CI 'presented an innovative way of addressing area-based urban challenges, effectively leading the way for a sea-change in thinking on urban regeneration in many member-states, both in terms of content and process' (Carpenter 2006).

Also noteworthy is the growing recognition from the EC of the importance of the role that cites should play to promote the process of territorial development within the EU, which was crystallised in 2012, when the 'Directorate General for Regional Policy' was renamed 'Directorate General for Regional and Urban Policy', and when at least 50% of EU Cohesion Policy (2014–20) funding was directly allocated to EU urban areas. Furthermore, a growing number of voices argue that the conception and implementation of EU policies need to be adapted to urban realities, and that cities need to be adequately involved in these processes. Alongside, 'an increasing number of sectoral EU policies explicitly target urban areas: Energy, Information Society, Environment, Climate Action, Education and Culture, Transport, etc. support initiatives such as European Capital of Culture, Smart Cities and Communities European Innovation Partnership, Green Capital Award, Covenant of Mayors and Mayors Adapt' (EC 2014b: 7).

At an initial phase, the main concerns of the EU towards urban-related policy measures were centred in reducing the excessive concentration of activities and inhabitants in the most urbanised areas of the EU, and in contributing to urban renewal and physical regeneration of decaying industrial sites. At the same time, in 1994, 'the Commission launched the URBAN Initiative, aimed at social and economic regeneration of cities and at improving the environment. URBAN has sought to maximise the involvement of the grassroots, empowering local communities and encouraging local people to determine priorities and to take responsibility for their own areas' (EC 1996: 110).

As time went by, social-related concerns were added to the EU Urban Policy measures, with the goal of mitigating existing high levels of poverty and exclusion within EU urban areas. Again, the EC recognised the importance of cities as key locations for the pursuit of a strategy for cohesion and sustainable development, and as economic centres for the development of the surrounding suburban and rural areas. Finally, it postulated that 'networks of large cities can stimulate a more balanced and polycentric form of development in which medium-sized towns and cities can play a key role' (EC 2001: 3), following the ESDP rationale to promote a more polycentric territory.

By the mid-2000s, the URBAN CI was on its second phase, and covered 44% of the EU population who lived in urban areas, with over 50,000 inhabitants. In general, this CI 'has focused, in particular, on creating and improving local social capital, in part by including active learning measures as an integral part of programmes'. Moreover, it 'acted as a catalyst for regeneration and, in some cases, has had a major leverage effect on investment' (EC 2004: 159).

For the 2007–2013 EU Cohesion policy programming period, the EC developed, in cooperation with the European Investment Bank (EIB) and the Council of Europe Development Bank (CEB), the Joint European Support for Sustainable Investment in City Areas, an innovative financial instrument better known as JESSICA, to promote urban development in the following thematic areas[1]:

- Urban infrastructure: including transport, water/wastewater, energy;
- Heritage or cultural sites: for tourism or other sustainable uses;
- Redevelopment of brownfield sites: including site clearance and decontamination;
- Creation of new commercial floor space for SMEs, IT and/or R&D sectors;
- University buildings: medical, biotech and other specialised facilities;
- Energy efficiency improvements.

In addition, concerns were placed on the environmental problems derived by the growth of suburbanisation processes, and the consequent decline of city centres, 'with shops and other businesses closing down. This calls for effective management of land-use and public transport as well inner-city renewal to slow down or even reverse the trend' (EC 2007: 53). Also, the JESSICA instrument has contributed to the revitalisation, modernisation and adaptation of urban areas with a view for them to become more sustainable over time (Tarnawska and Rosiek 2015). It has been found, however, that despite all EC efforts to elevate the urban dimension of EU policies, 'most cities have so far had a limited role in policy design and implementation and there are few signs of active participation of local residents' (EC 2010: 234). This EC financial assistance for projects is offered in the form of loans, guarantees or equity for urban investment projects, which are to be implemented by public or private investors or in public–private partnerships, thus allowing for leveraging additional financial support (Dąbrowski 2014). On a positive note, JESSICA provides a sustainability and recyclability logic, by ensuring 'that resources will be reinvested in a constant way,

[1] http://ec.europa.eu/regional_policy/en/funding/special-support-instruments/jessica/#2

to facilitate the implementation of projects aiming to sustainable urban development' (Patlitzianas 2011: 371).

For the 2014–2020 policy phase, the EC will direct financial support for sustainable urban mobility, regeneration of deprived communities and improved research and innovation capacity in urban areas, whilst a minimum of 5% of the European Regional Development Fund (ERDF) is earmarked for Integrated Sustainable Urban Development. Moreover, an urban development network (UDN) will be created in order to review on-the-ground deployment of European funds and the boosting knowledge-sharing performance between cities involved in integrated sustainable urban development and in Urban Innovative Actions. Finally, the URBACT III programme, acting as a European exchange and learning programme for promoting sustainable urban development, was financially strengthened and expanded, thus further enabling European cities to work together to develop better solutions to urban challenges. In synthesis, this programme has acted as an amplifier for participatory approaches with 'practitioners, city managers, elected representatives and stakeholders from other public agencies, the private sector and civil society' (URBACT 2014: 9).

Curiously, a more 'neoliberalist' vis-a-vis 'cohesive' vision for EU urban development is envisaged in the sixth EU Cohesion Report, which claims that the productivity of cities increases with its size, as larger cities are endowed with higher levels of human capital, larger share of high productivity sectors, 'greater supply of local public goods, as well as "shared", or common, facilities such as public laboratories and universities' (EC 2014a: 19) thus profiting from 'agglomeration benefits'. In a sense, the 'polycentric and balanced urban rationale' seemed to have given way to a 'growth and competitiveness rationale', very much coined with financial crisis eras.

On the most recent (seventh) EU Cohesion Report, however, the focus on the need to promote sustainable cities regained the EC attention, which highlighted the cities' potential solutions to current environmental challenges: 'while urban areas in the EU generally face more environmental challenges than other places, they can often prove to be more resource and energy-efficient than other areas where low density settlements, energy-intensive buildings (e.g. detached houses) and the level of dependency on the car for transport are generally more common. Housing in cities tends not only to occupy less land but also more frequently takes the form of apartments and townhouses, which generally require less energy to heat and cool' (EC 2017: 110). Moreover, EU Cohesion Policy started to pay particular attention to the specific socio-economic characteristics of EU functional areas and the promotion of the already mentioned Integrated Urban Development Strategies (van der Zwet et al. 2017).

In parallel, the ESPON programme supports a vision for a more open and polycentric European territory towards 2050, as increasing urbanisation levels and regional disparities are expected to occur in the following decades. Worse still, rising urbanisation is expected to lead, in many places, to uncontrolled urban sprawl processes. In this context, 'to improve its Territorial Cohesion, Europe needs to become more open and polycentric, fulfilling the original aim of the Treaty of Rome (1956) saying that Europe has to become an open Community of equals with common strong institutions, and as well the aim of later Treaties to opt for a harmonious and balanced territory' (ESPON 2014b: 12).

1.3 The Rising of the EU Urban Agenda

The idea to forge the elaboration of a dedicated EU Urban Agenda is not new. Indeed, by 1997, the EC sponsored a Communication 'Towards an urban agenda in the European Union'. Later on, by 2011, the 'European Parliament adopted a resolution arguing for a strengthening of the urban dimension of EU policies and the intergovernmental co-operation on urban development policies, calling for a joint working programme or European Urban Agenda' (EC 2014b: 8). Reflecting the same preoccupations, by 2013 the Committee of the Regions (CoR) launched its own initiative opinion entitled 'Towards an integrated urban agenda for the EU' (EC 2014b).

In sum, the urban question has become a key focus and an increasingly important issue within EU policies in recent years (Partidário and Correia 2004). It is also worth underlying that, since the first document adopted by the EC related to urban and territorial development, known as the Leipzig Charter on Sustainable European Cities (2007), was published, the initial focus placed on solving issues related with deprived neighbourhoods, within the context of a city, has shifted towards the need to promote a smarter, more sustainable and socially inclusive urban development, during the Informal Ministerial Meeting on Urban Development Declaration, which took place in Toledo, in 2010.

After several other EU initiatives, which discussed urban issues on a formal level, the Pact of Amsterdam, also known as the 'Urban Agenda for the EU', was adopted in 2016, during the informal meeting of EU ministers responsible for urban development. By providing objectives, thematic priorities, actions and operational frameworks, this 'Agenda' highlights 12 thematic priorities, and 'creates a new model of multilevel and multidimensional cooperation for urban policy stakeholders, whose aim is to strengthen the urban dimension in European Union policies' (Olejnik 2017: 177–178).

In a brief overview, however, the EC still lacks the legislative powers, financial resources, organisational capacity and political power to develop a dedicated 'EU Urban Policy'. As such, the assistance given by the EC to urban issues will continue to be based on supporting 'urban policy experimentation, dissemination of best practice and a gradual raising of urban issues on the policy agenda' (Atkinson 2001: 399). In a different prism, the future of the EU Urban Policy will depend on specific crisis conditions, the policy integration/disintegration (read Brexit) trends and the empowerment and resource capacity of urban/regional authorities, amongst other factors (Delladetsima 2003).

Furthermore, the re-orientation of the EU structural and investment funds for the 2021–2027 programming period would certainly affect EC initiatives directly involved in urban policy affairs, as occurred in previous programming periods (Lang and Török 2017). In this regard, EU Cohesion Policy is of critical importance to finance Urban Development Policies as it is viewed as 'one of, if not, the largest integrated development policies in the Western world, and one of the largest of such programmes anywhere in the world' (Mccann and Vargam 2015: 1255). It is also crucial to point out that these increasing political concerns over the 'urban question' are not exclusive of the EU. In the bigger picture, the United Nations have recently (2016) defined 17 Sustainable Development Goals (SDG), one having an exclusive urban focus: Goal 11—Make cities and human settlements inclusive, safe, resilient and sustainable (Caprotti et al. 2017).

Regarding the EU territory, presently, the EU Urban Agenda seeks to improve the quality of life of citizens in urban areas, in a new working method to ensure maximum utilisation of the growth potential of cities and to successfully tackle social challenges. More concretely, 'partnerships are set-up around 12 priority themes (air quality, circular economy, climate adaptation, digital transition, energy transition, housing, inclusion of migrants and refugees, innovative and responsible public procurement, jobs and skills in the local economy, sustainable use of land and nature-based solutions, urban mobility, urban poverty) with European and urban relevance. Within these Partnerships, problems will be identified and solutions will be recommended through action plans (these are addressed to the EU, the Member States and the cities). The action plans will contain actions and also examples of good projects to be scaled-up and transferred across the EU. Actions could be a proposal to amend an EU Directive, or a proposal for the new European Structural and Investment Funds (ESIF) reinforcing cooperation on shared issues (…)'.[2] In a complementary way, the URBACT Programme helps cities to develop pragmatic solutions that are new and sustainable and that integrate economic, social inclusion, integrated urban development, environmental and governance urban-related topics[3] (Table 1.3).

Regarding the EU Urban Agenda strategic rationale, it follows the ESDP vision to promote a more polycentric and balanced development of the EU, with a view to achieving economic, social and territorial cohesion. Additionally, it supports the logic of an integrated urban policy approach. For this, the ESPON programme has developed a wealth of knowledge and evidence relating to the territorial dimension of the European urban fabric, since 2002 (ESPON 2014a). On top of this, the EU Urban Agenda 'is aimed at promoting cooperation between Member States, cities, the European Commission and other stakeholders in order to maximise the growth potential of cities and to tackle social problems and so to improve the quality of life in urban areas' EC 2017: 133). Crucially, it defined 12 priority themes and cross-cutting issues, with a widespread concern to current policy preoccupations, mostly related with the rise of migrant inflows, environmental, governance and urban mobility concerns (Urban Agenda 2016):

- Inclusion of migrants and refugees;
- Air quality;

---

[2] http://ec.europa.eu/regional_policy/en/policy/themes/urban-development/

[3] http://urbact.eu/

Table 1.3 URBACT intervention areas

Urban Dimension	Component
Environment	• Abandoned Spaces • Energy Efficiency • Food • Housing • Low Carbon • Risk Prevention • Urban Mobility • Waste
Governance	• Capacity Building • City Branding • City Management • City Planning • Cross-border Cooperation • Financial Engineering • Knowledge Economy • Participation • Social Innovation • Urban–rural • Youth
Integrated Urban Development	• Abandoned Spaces • Capacity Building • City Planning • Culture and Heritage • Disadvantaged Neighbourhoods • Energy Efficiency • Financial Engineering • Housing • Low Carbon • Strategic Planning • Urban Mobility • Urban Renewal • Urban Sprawl • Urban–rural
Economy	• Circular Economies • City Branding • Culture and Heritage • Employment • Entrepreneurship and SMEs • Food • Health • Knowledge Economy • Local Economic Development • Research and Innovation • Sharing Economy • Waste

(continued)

Table 1.3 (continued)

Urban Dimension	Component
Inclusion	• Ageing • Culture and Heritage • Disadvantaged Neighbourhoods • Education • Employment • Health • Housing • Migrants • Minorities • Participation • Poverty • Roma • Sharing Economy • Social Innovation • Youth

Source Own elaboration

- Urban poverty;
- Housing;
- Circular economy;
- Jobs and skills in the local economy;
- Climate adaptation (including green infrastructure solutions);
- Energy transition;
- Sustainable use of land and nature-based solutions;
- Urban mobility;
- Digital transition; and
- Innovative and responsible public procurement.

In essence, the EU Urban Agenda aims at strengthening the urban dimension of EU policies, and supporting a greater coherence between urban and regional development agendas, whilst fostering the urban dimension in the context of Territorial Cohesion. From this general perspective, it builds on the Leipzig Charter's (2007) prerogative of making greater use of Integrated and Sustainable Urban Development Policy approaches.

1.4 Recommendations for Cohesive Urban Policies—Conclusive Remarks

As Storper and Scott (2016: 1114) claim, 'the current period of human history can plausibly be identified not only as a global but also as an urban era. This is a period in which population, productive activity and wealth are highly and increasingly concentrated in cities'. As previously explained, the concept of Territorial Cohesion builds bridges between several territorial development dimensions which include, in our understanding, a marked urban dimension, which we designate polycentrism.

Indeed, according to the latest Territorial Agenda (2011: 4), 'territorial cohesion is a set of principles for harmonious, balanced, efficient, sustainable territorial development'. Europe being a markedly urbanised territory, this development goal requires an 'urban development focus'. Likewise, worldwide, more than 50% of human beings already dwell in urban areas, and this trend is on the rising. These potential 'Territorial Cohesion Urban Development Policies' should promote a more polycentric pattern vis-à-vis a concentrated/monocentric one, in order to 'act as centres contributing to the development of their wider regions' (Territorial Agenda 2011: 5). One way of achieving this is to concentrate public investment in medium towns located in less developed areas, in an urban development strategic rationale coined as 'Territorial Cohesion Cities' (Medeiros and Rauhut 2018).

In a complementary way, territorial cohesion processes will benefit from encouraging urban integrated and multilevel development strategies, which take into account all territorial levels in territorial planning instruments, and development policies. Associated with this policy rationale is the concept of functional regions, which aim to better explore territorial complementarities and to tackle common territorial needs, beyond a large or medium-city administrative border, whilst contributing to a wider integration of their peri-urban neighbourhoods, and/or peripheral rural regions. At the same time, urban integrated development strategies should support cross-border and transnational functional regions, as a means to reducing border barriers and to promoting cross-border and transnational planning, geared towards the Territorial Cohesion policies' main goals.

Understandably, the world being more and more urbanised, the contribution of urban development policies tends to gain increasing critical importance to the achieving of Territorial Cohesion trends worldwide. For that, there is a need to strengthen potential synergies between sustainable urban development, social integration, innovative economy, territorial connectivity, and multilevel and sound urban governance and planning processes. The key challenge here is to ensure a more balanced and polycentric urban pattern across territories, in order to avoid excessive concentration of people and activities in a few very large urban agglomerations, which favours the occurrence of negative environmental (pollution) and social (criminality, poverty) impacts, whilst limiting the potential to better exploring the territorial capital of vast depopulated territories.

From a governance point of view, the coordination between the local- and city-regional levels should be strengthened, as well as the partnerships between larger, medium/small cities and rural areas (Leipzig Charter 2007), as a means to boost territorial efficiency. This prevailing vision in which 'there is a need to enhance the complementarity of policies affecting urban areas and to strengthen their urban dimension' (Urban Agenda 2016: 4) needs to be considered by the metropolitan, regional and national planning instruments, as the main tool to achieving Territorial Cohesion processes. Moreover, urban authorities should increase their cooperation with stakeholders, including the ones related with civil society, knowledge institutions and local communities.

On a critical note, whilst the EU Urban Agenda acknowledges the need to support urban areas of all sizes, we propose a concentration of public development investment in the already mentioned 'Territorial Cohesion Cities', a concrete policy measure to achieve Territorial Cohesion processes. Understandably, the complexity of urban challenges requires a tailor-made approach to the characteristics of each territory. Moreover, the need to promote sound and strategic urban planning processes in the EU Urban Agenda is not associated with the prerequisite to reducing urban sprawl processes vis-à-vis a more compact, planned and efficient urban morphology vision.

In all, we can conclude that urban development policies are gaining a major role in achieving Territorial Cohesion processes, with increasing worldwide urbanisation processes. For this goal to become a reality, however, cities located on less developed areas should experience higher development levels when compared with the ones located in more developed areas, in a baseline scenario. Thus, since medium towns, or second-tier cities, have the potential to be more efficient in making the most of the regional territorial capital than smaller urban areas, we suggest that available EU and national public funding should be concentrated for their development, thus contributing to achieving a more polycentric, balanced and harmonious development, towards the EU Territorial Cohesion policy goal.

References

Abrahams G (2014) What "Is" Territorial Cohesion? What does it "Do"? Essentialist versus pragmatic approaches to using concepts. Eur Plan Stud 22(10):2134–2155. https://doi.org/10.1080/09654313.2013.819838

Atkinson R (2001) The emerging 'Urban Agenda' and the European spatial development perspective: towards an EU urban policy? Eur Plan Stud 9(3):385–406

Bradley J, Zaucha J (2017) Territorial cohesion: a missing link between economic growth and welfare Lessons from the Baltic Tiger. Uniwersytet Gdański Katedra Makroekonomii, Gdansk

Caprotti F, Cowley R, Datta A, Broto VC, Gao E, Georgeson L, Herrick C, Odendaal N, Joss S (2017) The new urban agenda: key opportunities and challenges for policy and practice. Urban Res Pract 10(3):367–378

Carpenter J (2006) Addressing Europe's urban challenges: lessons from the EU URBAN community initiative. Urban Stud 43(12):2145–2162

Clifton J, Díaz-Fuentes D, Fernández-Gutiérrez M (2016) Public infrastructure services in the European Union: challenges for territorial cohesion. Reg Stud 50(2):358–373. https://doi.org/10.1080/00343404.2015.1044958

Colomb C, Santinha G (2014) European Union competition policy and the European territorial cohesion agenda: an impossible reconciliation? State aid rules and public service liberalization through the European spatial planning lens. Eur Plan Stud 22(3):459–480. https://doi.org/10.1080/09654313.2012.744384

COR (2003) Territorial cohesion in Europe. Committee of the Regions, Brussels

Chorianopoulos I, Iossifides T (2006) The neoliberal framework of EU urban policy in action: supporting competitiveness and reaping disparities. Local Econ 21(4):409–422

Dąbrowski M (2014) Engineering multilevel governance? Joint European support for sustainable investment in city areas (JESSICA) and the involvement of private and financial actors in urban development policy, regional studies. Reg Stud 48(12):2006–2019

Dao H, Cantoreggi PP, Rousseaux V (2017) Operationalizing a contested concept: indicators of territorial cohesion. Eur Plan Stud 25(4):638–660. https://doi.org/10.1080/09654313.2017.1281230
Davoudi S (2005) Understanding territorial cohesion. Planning Pract Res 20(4):433–441
Delladetsima P (2003) What prospects for urban policy in Europe? City 7(2):153–165
EC (1996) First report on economic and social cohesion. European Commission, Luxemburg
EC (1999) ESDP. European spatial development perspective. Towards balanced and sustainable development of the territory of the European Union, European Communities, Luxemburg
EC (2001) Second report on economic and social cohesion. European Commission, Luxemburg
EC (2004) Third report on economic and social cohesion, Convergence. Competitiveness and cooperation. European Commission, Luxemburg
EC (2007) Fourth report on economic and social cohesion, growing regions, growing Europe. European Commission, Luxemburg
EC (2008) Green paper on territorial cohesion. Turning territorial diversity into strength. Communication from the Commission to the Council, The European Parliament, The Committee of the Regions and the European Economic and Social Committee, Oct 2008, Brussels
EC (2010) Investing in Europe's future: fifth report on economic, social and territorial cohesion. European Commission, Nov 2010, Brussels
EC (2014a) Sixth report on economic, social and territorial cohesion. Investment for jobs and growth: promoting development and good governance in EU regions and cities, European Commission, July 2014, Brussels
EC (2014b) The urban dimension of EU policies—key features of an EU urban agenda. COM (2014) 490 final, European Commission, Brussels
EC (2017) Seventh report on economic, social and territorial cohesion. My Region, My Europe, Our Future. European Commission, Brussels
EEA (2010) The territorial dimension of environmental sustainability. Potential territorial indicators to support the environmental dimension of territorial cohesion. European Environment Agency. EEA Technical Report no 92010, Copenhagen
ESPON 3.2 (2006) Spatial scenarios and orientations in relation to the ESDP and cohesion policy, ESPON project 3.2, Third Interim Report. January 2006, Luxembourg
ESPON INTERCO (2011) Indicators of territorial cohesion, Scientific Platform and Tools Project 2013/3/2, (Draft) Final Report, Part C, Scientific report, ESPON & University of Geneva, Luxembourg
ESPON KITCASP (2012) Key indicators for territorial cohesion and spatial planning, Interim Report, Version 31/10/2012, Luxembourg
ESPON (2014a) Territorial dynamics in Europe. Evidence for a European urban agenda, Territorial Observation no 13, European Spatial Planning Observatory Network, Luxembourg
ESPON (2014b) Making Europe open and polycentric. Vision and scenarios for the European territory towards 2050, European Spatial Planning Observatory Network, Luxembourg
ESPON (2016) ESPON policy briefing—polycentric territorial structures and territorial cooperation. European Spatial Planning Observatory Network, Luxembourg
Faludi A (2004) Territorial cohesion: old (French) wine in new bottles? Urban Stud 41(7):1349–1363
Faludi A (2005) Territorial cohesion: an unidentified political objective: introduction to the special issue. Town Planning Rev 76(1):1–13. https://doi.org/10.3828/tpr.76.1.1
Faludi A (2006) From European spatial development to territorial cohesion policy. Reg Stud 40(6):667–678. https://doi.org/10.1080/00343400600868937
Faludi A (2007) Territorial cohesion policy and the European model of society. Eur Plan Stud 15(4):567–583. https://doi.org/10.1080/09654310701232079
Glinka K (2017) Integrated territorial investment as instrument for managing transport security in lower Silesia's largest cities. Sci J Silesian Univ Technol Ser Transp 94(2017):37–46
González A, Daly G, Pinch P, Adams N, Valtenbergs V, Burns MC, Johannesson H (2015) Indicators for spatial planning and territorial cohesion: stakeholder-driven selection approach for improving usability at regional and local levels. Reg Stud 49(9):1588–1602. https://doi.org/10.1080/00343404.2015.1018883

Holder J, Layard A (2011) Drawing out the elements of territorial cohesion: re-scaling EU spatial governance. Yearb Eur Law 30(1):358–380. https://doi.org/10.1093/yel/yer016

Janin Rivolin U (2005) The future of the ESDP in the framework of territorial cohesion, disP—The Planning Review 41(161): 19–27. https://doi.org/10.1080/02513625.2005.10556921

Kamrowska-Zaluska D, Obracht-Prondzynska H (2017) Implementation of the integrated territorial investments. In: Medeiros E (ed) Uncovering the territorial dimension of European Union cohesion policy. Routledge, London and New York, pp 114–126

Lang T, Török I (2017) Metropolitan region policies in the European Union: following national, European or neoliberal agendas? Int Plan Stud 22(1):1–13

Leipzig Charter (2007) Leipzig Charter on sustainable European cities, agreed on the occasion of the Informal Ministerial Meeting on Urban Development and Territorial Cohesion in Leipzig on 24/25 May 2007, Leipzig

Luukkonen J, Moilanen H (2012) Territoriality in the strategies and practices of the territorial cohesion policy of the European Union: territorial challenges in implementing "soft planning". Eur Plan Stud 20(3):481–500. https://doi.org/10.1080/09654313.2012.651806

Mccann P, Vargam A (2015) The reforms to the regional and urban policy of the European Union: EU cohesion policy. Reg Stud 49(8):1255–1257

Medeiros E (2005) A coesão territorial nas NUTS III de fronteira de Portugal Continental—A iniciativa INTERREG-A e o desenvolvimento regional. O caso da NUT III do Alentejo Central, Master Thesis, Lisbon University, Lisbon

Medeiros E (2014) Territorial cohesion trends in inner Scandinavia: the role of cross-border cooperation (INTERREG-A 1994–2010). Nor Geogr Tidsskr 68(5):310–317. https://doi.org/10.1080/00291951.2014.960949

Medeiros E (2016a) Territorial cohesion: an EU concept. Eur J Spat Dev 60. Available from: http://www.nordregio.se/Global/EJSD/Refereed%20articles/refereed60.pdf

Medeiros E, Rauhut D (2018) Is territorial cohesion possible via the EU cohesion policy? Testing a 'territorial cohesion cities' policy rationale, Regional Studies, being reviewed

Nijkamp P, Kourtit K (2013) The "New Urban Europe": global challenges and local responses in the urban century. Eur Plan Stud 21(3):291–315

Olejnik A (2017) Future perspectives of the implementation of EU urban Agenda. Interdisc Polit Cult J 19(1):175–181

Partidário MR, Correia FN (2004) Polis—the Portuguese programme on urban environment. A contribution to the discussion on European Urban Policy. Eur Plan Stud 12(3):409–423

Patlitzianas K (2011) An analysis of energy efficiency investments' environment in Greece—the potential role of JESSICA instrument. Energy Convers Manage 52(2011):366–373

Satterthwaite D (2016) Successful, safe and sustainable cities: towards a New Urban Agenda. Commonwealth J Local Governance 19(2016):3–18

Schön P (2005) Territorial cohesion in Europe? Plan Theory Pract 6(3):389–400. https://doi.org/10.1080/14649350500209397

Servillo L (2010) Territorial cohesion discourses: hegemonic strategic concepts in European spatial planning. Plan Theory Pract 11(3):397–416. https://doi.org/10.1080/14649357.2010.500135

Storper M, Scott AJ (2016) Current debates in urban theory: a critical assessment. Urban Stud 53(6):1114–1136

Tarnawska K, Rosiek J (2015) The JESSICA initiative: an instrument for urban sustainable development. Examples of urban regeneration in Silesia (Poland) and Central Moravia (Czech Republic). Comp Econ Res 18(2):119–138

Territorial Agenda 2020 (2011) Territorial agenda of the European Union 2020—towards an inclusive, smart and sustainable Europe of diverse regions. Retrieved from http://www.eu2011.hu/files/bveu/documents/TA2020.pdf

URBACT (2014) URBACT III operational programme. European regional development fund 2014–2020, European Territorial Cooperation, European Programme for Sustainable Urban Development, Brussels

Urban Agenda (2016) Establishing the Urban Agenda for the EU. 'Pact of Amsterdam'. Agreed at the Informal Meeting of EU Ministers Responsible for Urban Matters on 30 May 2016 in Amsterdam, The Netherlands, Amsterdam

van der Zwet A, Bachtler J, Ferry M, McMaster I, Miller S (2017) Integrated territorial and urban strategies: how are ESIF adding value in 2014–2020? Final Report, Directorate-General for Regional and Urban Policy, European Commission

Van Well L (2012) Conceptualizing the logics of territorial cohesion. Eur Plan Stud 20(9):1549–1567. https://doi.org/10.1080/09654313.2012.708021

Vanolo A (2010) European spatial planning between competitiveness and territorial cohesion: shadows of neo-liberalism. Eur Plan Stud 18(8):1301–1315. https://doi.org/10.1080/09654311003791358

Zillmer S, Böhme K, Lüer C (2012) Territorial cohesion in future EU cohesion policy. Final report for the research project "The territorial dimension of future EU cohesion policy" (Forschung issue 156 ed, p 53). Federal Ministry of Transport, Building and Urban Affairs (BMVBS), Berlin

Eduardo Medeiros is a Geography Professor and a Senior Research Integrated Fellow in DINÂMIA'CET-IUL, Instituto Universitário de Lisboa (ISCTE-IUL), Portugal. He has a Ph.D. in Geography—Regional and Urban Planning, and has more than 30 published papers in international journals, eight books and nine book chapters. His research interests are focused on Territorial Cooperation, Territorial Cohesion, Territorial Development and Territorial Impact Assessment. He is a DG REGIO (European Commission) expert and a Horizon 2020 evaluator. He is also a Regional Studies Association fellow and belongs to its Cohesion Policy Research Network. He has coordinated several international policy evaluation projects and is a member of an ESPON project. He was invited as a project adviser and to write reports by DG REGIO. He was already invited to be a keynote speaker by several International Universities and EU institutions (European Commission and Committee of the Regions). He is part of the scientific and editorial committee of some journals and a peer reviewer of more than 20 international journals.

Chapter 2
Civil Society in Urban Areas: A Partner for Territorial Cohesion?

Oto Potluka and Martin Špaček

Abstract Territorial dimension and integrated approaches to regional and urban development became an important issue in programmes financed by the EU Cohesion Policy. Local actors and nonprofit organisations (NPOs), as the crucial representatives of civil society, belong to partners in creation and implementation of EU integrated approaches, such as integrated urban development plans. However, the participative approaches to governance have been developing slowly in Central and Eastern Europe. The Czech Republic's case in this chapter shows how the NPOs with their limited capacities tackle with such circumstances and how NPOs can influence territorial cohesion processes. For this reason, we investigate the role of Czech civil society in territorial cohesion, particularly in Integrated Urban Development Plans implemented in programming period 2007–2013.

Keywords Urban areas · Territorial cohesion · Civil society · NPOs
Integrated urban development

2.1 Introduction

Good living conditions lead to greater satisfaction. In such conditions, people have a feeling of belonging to a community, which helps form cohesion in the city in which they live. This contributes to a stable or growing population in cities because people are less likely to move away. The conditions for such a situation are based on a combination of the city as both a pool of business activities and a place to live (Porter 2008, pp. 373–405). It is not an easy process, as city governance must take into account the fact that global economic pressure affects city development

O. Potluka (✉)
Centre for Philanthropy, University of Basel, Basel, Switzerland
e-mail: oto.potluka@unibas.ch

M. Špaček
CETIP Network, Prague, Czech Republic
e-mail: spacek@e-academia.eu

E. Medeiros (ed.), *Territorial Cohesion*, The Urban Book Series,
https://doi.org/10.1007/978-3-030-03386-6_2

(Keil 2011) and people are willing to shape the place where they live; otherwise, social inequality and economic instability may intensify (Novy et al. 2012). From this perspective, urban development needs to be territorially balanced and socially cohesive.

Some parts of cities have strong potential to become poles of growth, while others are confronted with social problems. This imbalance might cause problems in social and territorial cohesion (Trudelle et al. 2015). Districts lagging behind in their development need both bridging and bonding social capital, while developed areas are more equipped with bonding social capital and do not need to add bridging social capital (Putnam 2000). Although European cities represent typical cases of bonding capital (Vidoni et al. 2009), they also cope with certain difficulties. To tackle these difficulties, city political representatives present development plans supported by the EU Cohesion Policy as integrated urban development plans (IUDPs). Such plans are designed to simultaneously contribute to meeting the objectives of good living conditions and economic development and thus increase both bridging social capital among different city districts and bonding social capital within the city as a whole. These plans are usually oriented towards investment in tangible assets, as the development concerns primarily economic goals (Mascherini et al. 2011), while social cohesion has both physical and moral dimensions.

The European Union fosters the participation of civil society organisations regarding their inclusion in local planning (Andreotti et al. 2012). This top-down approach raises several questions: Do these plans contribute to social cohesion, and if so, how? What is the role of local initiatives such as nonprofit organisations and civic engagement in the design and implementation of such plans? Are these plans and programmes sustainable? If IUDPs are successful, population size should not change or should grow in areas with IUDPs. Still, some questions remain. If the outcomes of such policies have impacts on these areas, are people capable of influencing these policies? Are inhabitants the objects or the subjects of such plans? Our chapter contributes to the discussion on city cohesion by answering these questions.

To that end, we present research on local actors and nonprofit organisations (NPOs) as crucial representatives of civil society, belonging to partners in the creation and implementation of EU integrated approaches, such as the IUDPs implemented in the period 2007–2013. Integrated territorial investments (ITIs) and Community-led local development (CLLD) are instruments still being implemented in the period 2014–2020; thus, we concentrate our research on the instruments that have already been implemented.

The remainder of the chapter is organised as follows. First, we introduce conceptual considerations of the contribution of nonprofit organisations to the production of public goods, social and territorial cohesion, and the expected effects of co-production on them. Moreover, we discuss EU instruments for urban development. Second, we discuss the Czech experience with integrated approaches and the partnership principle at a local level, with the analysis showing low actual social and political participation in city development. The final section concludes.

2.2 Contribution of Civil Society to Territorial Cohesion

In discussing the contribution of civil society and civic engagement to territorial cohesion, we need to take into account three factors. The first factor concerns the political system in which civil society adds to territorial cohesion. Together with overall social trust, the political system influences civic engagement and nonprofit organisations' activities in policymaking (both formal and informal). This makes civic engagement in territorial cohesion less possible in countries and places with low social capital. The second factor relates to the question of what types of goods and services the nonprofit organisations provide or should provide. The third factor concerns the co-creation and co-production of public services, partnerships and civic engagement. While the first two are more macro-level issues, the third is a local-level issue.

2.2.1 Local Political System, Social Capital and Trust in Nonprofit Organisations in the Czech Republic

People can articulate their engagement in public affairs not only by casting ballots in elections but also by gathering in nonprofit organisations or personally engaging in policymaking. These means also enable them to communicate their needs to politicians between elections. Politicians should hear these voices to make policies that are better fitted to these needs and that are thus sustainable (EC 2004). Moreover, such bilateral communication helps increase social territorial cohesion.

Central and Eastern Europeans express scepticism towards and dissatisfaction with democracy (18%) and its performance (62%) (Karp and Milazzo 2015). Thus, political parties lose their positions, and bottom-up political movements benefit as a result. Such a situation is projected not only at the parliamentary level but also even more strongly at the local level, especially in cities. As Potluka et al. (2018, forthcoming) point out for the Czech Republic, 43.0% of local election candidates were from parliamentary political parties in 2010, while only 35.3% were in 2014. Among the elected candidates in 2010 (2014), 30.4% (23.5%) were standing for a seat on lists of parliamentary political parties in the Czech Republic. Thus, independent candidates constitute a major part of local assemblies. The situation also differs in rural and urban areas. In cities, especially in the capital city, party politics still play an important role (Potluka and Perez 2018 forthcoming). This development reflects the fact that people trust local politicians who are closer to them, know their needs and help create better places to live.

In post-communist countries, general trust in other people is quite low. In the Czech Republic, only approximately one-fifth of the population express that most people can be trusted or that they cannot be too careful (ESS Round 7: European Social Survey Round 7 Data 2014). Moreover, general distrust is echoed in disbelief in nonprofit organisations and their capability to influence policies and be effective political agents (EC 2013). More than half of the Czech population (52%) perceives that membership in a nonprofit organisation is an ineffective way to influence political life (EC 2013).

Low financial capacity, insufficient advocacy, post-communism mistrust in organisations and missing friendship networks cause this situation (Frič 2004; Howard 2011; Rose-Ackerman 2007). Moreover, the role of civil society in political life was not established because of long-term disputes among political elites (Potůček 1999).

2.2.2 *Provision of Public Goods in Cities by the Nonprofit Sector*

People want to live in an environment with good living conditions. Their expectations lead to massive migration to cities. In Europe, the share of the urban population compared to the total population is projected to increase from 73% in 2014, to 82% in 2050 (UN 2015, p. 50). Such a massive migration and population living in a city exert pressure on city infrastructure because of high population densities. This pressure then causes the loss of social and territorial cohesion and necessitates additional investments in infrastructure on both the city and regional levels. This raises several questions. Who should be responsible for the production of such conditions? Is it the responsibility of the public or private sector to build infrastructure? The literature gives an equivocal answer to these questions of the production of collective goods.

The traditional roles of sectors in the production of collective goods in mixed markets relate to both market and government failures (Hansmann 1980; Weisbrod 1975). Three types of market failures appear: (i) the under-provision of goods and services, (ii) the over-exclusion of consumers and (iii) contract failure. The first type of market failure is caused by the fact that market forces leave some segments of demand unsatisfied due to demand heterogeneity [for example, because of demographic heterogeneity (Andreotti et al. 2012)]. In the second case, some people are excluded from the consumption of private goods and services by others consuming the same goods and services. The third type of market failure arises in cases of information asymmetry where suppliers tend to exploit an information advantage (i.e. better knowledge of the quality of delivered goods or services) that they have over their customers.

Local governments react to these market failures by providing collective goods, buying goods from firms and the nonprofit sector, regulating the market or providing selective subsidies. The public sector, however, fails to satisfy some components of demand even at the local level, as it provides goods and services according to the majority rule and the preferences of a median voter (Ben-Ner and van Hoomissen 1992; Weisbrod 1975) as well as limited budgets. Thus, in the case of heterogeneous preferences, unsatisfied demand is still present.

The failure of the market and the government gives rise to the existence of the nonprofit sector (hereafter referred to as NPOs, nonprofit organisations). NPOs could solve under-provision and over-exclusion by collecting donations and voluntary work to provide services for demand in excess of mainstream preferences, providing collective goods, and cross-subsidising. Market and government failures give NPOs a chance to deliver a specific quality and quantity of goods to satisfy particular segments of demand. This also concerns the provision of city services and includes activities and initiatives provided by voluntary neighbourhood initiatives and member-oriented NPOs, and it makes room for civic engagement. Thus, the leading position on local development policy is shifting from technocrats to individuals with appropriate capabilities (Bowden and Liddle 2018), widening the partners involved, including non-profit leaders.

2.2.3 *Role of Civil Society and Civic Engagement in City Planning and Development*

The provision of local public goods and city planning differ in how the local government allows local actors to take part. This can occur either as managed participation, when people choose only the specified options (Su 2018), or people and civil society may be active partners of the public sector in the co-production of collective/public goods and services (Vamstad 2012). Being in the position of consumers, not co-producers, people are not highly motivated to take part in the shaping of the city. Cities become more cohesive as a territorial identity is built and people feel like they belong to a group (Capello 2018). The similarity in interests and the solidarity within a group make the territory socially cohesive.

Members of socially cohesive local communities take care of the place in which they live. They are willing to participate in the design and implementation of policies that shape their living environment. They have an advantage in terms of the knowledge of local needs, which is reflected in political demand. Thus, the coordination of private decisions and public policies lead to the production of public goods (co-production).

However, low skills and low technical expertise limit such involvement (Andreassen et al. 2014; Su 2018; Vamstad 2012), or conflicts of interest appear (Lavasseur 2018). For a stronger voice, informal civic engagement transforms into formal structures, usually into organised civil society and NPOs.

Civil societies and civic engagement play an important role in Western European cities (Cassiers and Kesteloot 2012; Luria et al. 2015). Activating city residents to participate in political life and the community is an important tool for encouraging urban development projects (Paarlberg and Yoshioka 2016; Strom 2008). Even in countries with a long tradition of participation, the process of partnership and civil society participation in local decision-making is not easy. Although local residents may participate in renewal projects, it is difficult for them to influence the final decisions because of customary rigidities [for example, see the case of Norway in Hanssen (2010)].

Nevertheless, local leadership has the power to change the existing governance structure, even if its efforts are hampered by internal tensions caused by the different capacities of stakeholders (Cornforth et al. 2015; Eizaguirre et al. 2012). NPOs could exercise an important role by including the public in processes from which they would normally be excluded and by disseminating information to them. In this role, NPOs could act as mediators between individuals and municipalities, as they represent a wide spectrum of opinions and communicate them to the public sector.

In Central European countries, the participatory experience is quite different. Low trust among stakeholders (Potluka et al. 2014, 2018, forthcoming) and low capacities in partnership processes among NPOs (Baun and Marek 2008; Nałęcz, et al. 2015; Potluka et al. 2017) weaken the application of participatory approaches to public policy design and implementation.

2.2.4 Co-creation and Co-production of Public Services, Partnership and Civic Engagement

In recent years, the active participation of people in providing collective goods and services has received increasing attention from researchers. Among them, Brandsen and Pestoff (2006) distinguish three types of active participation. First, in co-production, people produce collective goods or services. At this micro-level, people are also consumers of these goods and services. Second, when people produce collective goods and services in cooperation with local governments, the authors speak of co-management on the meso-level. Third, when people participate in the planning of public policies and services, they identify it as co-governance. In this type of co-creation, NPOs as associations play an important role, as this activity is at a macro-level.

The current state of the art of research on co-creation and co-production shows a fascinating process, but the outcome of the process is not closely explored (Voorberg et al. 2015). The levels of participation in decision-making and in the provision of public policies belong to the drivers of policy performance (Pestoff 2012), but at the level of city planning, the issue is omitted. Interactions among public policy providers and people may support the process of quality achievements (Vamstad 2012). Thus, the already implemented IUDPs provide us with an excellent opportunity to build on co-creation and co-production process research and explore their effects on territorial and social cohesion.

For successful co-creation, stakeholder relationships have to be reciprocal, transparent and represent joint values continuously interacted among stakeholders (Vallaster and von Wallpach 2018). In the IUDPs, the stakeholders expect to receive subjective value for participating (for example, reconstructed public spaces), but the issue is whether these relationships will be sustained in the long term or if they will exist only during the design and implementation of IUDPs.

2.3 EU Cohesion Policy and Integrated Approaches to Urban Development

The EU Cohesion Policy covers both the economically and socially underdeveloped territories (usually rural) and developed regions (usually metropolitan areas). In regions with unfavourable conditions, the focus of regional policies is to promote balanced regional development (social and economic cohesion) among EU regions, whereas in developed regions, the focus is to increase their global competitiveness. These seemingly contradictory objectives of the EU Cohesion Policy often meet in urban areas, which can integrate both goals in their development.

Integrated approaches to urban development were supported in the programming period 2007–2013 by the European Regional Development Fund (ERDF) and the EU initiative JESSICA. The European Commission started to support 'the development of participative, integrated and sustainable strategies to tackle the high concentration of economic, environmental and social problems affecting urban areas' (EU 2006, Art. 8). The idea was to integrate solutions to the diversified problems in urban areas into one strategy based on multi-source funding and participative approaches. Thus, the key terms in such policies are integration, participation, sustainability and urban development. The understanding of these terms and the perception of their interdependences differ across EU Member States (Colini and Tripodi 2010).

Significant differences among the Member States prevail in the practical implementation of integrated approaches (Colini and Tripodi 2010). In the old Member States, these approaches were known and used in urban development before they were mainstreamed in the EU Cohesion Policy; however, in the new Member States, these approaches represented new tools in regional development (Škorňa 2011). The strategies were supposed to be implemented through financial support from different sources, and in the new Member States, in particular, the support within integrated approaches remained based on sectoral investments (Szokolai 2012). Whereas the old Member States focussed on interventions in deprived urban areas to achieve internal cohesion, in the new Member States, the focus was on reducing structural imbalances through investments in basic infrastructure (Colini and Tripodi 2010).

Aside from the complexity of integrated territorial development approaches to address diversified challenges, the common aspect of these instruments is the strong emphasis on the participation of different actors coming from different governmental levels and sectors. Their participation should be present during the whole policy process, from the creation and design of priorities, their management, implementation and monitoring, to the evaluation process (EC 2014b). However, in the new EU Member States, there is strong decision-making power located in the national government, and non-participatory approaches prevail in the implementation of the EU Cohesion Policy (Dąbrowski et al. 2014; Potluka et al. 2017). Thus, the stakeholders criticise the low involvement of cities in designing particular operational programmes in direct relation to integrated approaches when cities become partners only at implementation phases (Colini and Tripodi 2010; Szokolai 2012).

The experience gained through the implementation of integrated urban development approaches was transformed into more systematic support for integrated territorial development in the EU Cohesion Policy in the current programming period 2014–2020. Different integrated tools are promoted: ITIs and CLLD. The ITIs are specific integrated territorial tools implemented at the local level based on multidimensional and place-specific approaches, which enable cross-cutting funding from several priorities across operational programmes. Its expected benefits are seen in better aggregated outcomes, multi-source funding (which secures stability), the empowerment of subregional actors, and the release of unused potential at subnational levels (EC 2014c). The CLLD is a specific integrated tool based on the LEADER approach, focussed on multi-sectoral local action units at the subregional level (EC 2014a). Integrated approaches are typical examples of newly introduced place-based policies when each Member State can flexibly change the scope and scale of its implementation. Based on preliminary results, in the majority of cases, integrated development represents a multi-sectoral, multi-partner and multi-fund tool that leads to institutional innovations (van der Zwet et al. 2017).

2.4 The Czech Experience with Integrated Approaches and the Partnership Principle at the Local Level

Integrated approaches have begun to be implemented to a greater extent in the Czech Republic since its accession to the EU. In the programming period 2007–2013, these approaches were implemented in the form of IUDPs, which were financed from two sources: the Integrated Operational Programme (IOP) and eight Regional Operational programmes (ROPs). Only cities of over 50 thousand inhabitants for the ROPs (and the city of Mladá Boleslav) or over 20 thousand for the IOP can participate in these schemes. Altogether, 41 cities participated in IUDPs in the IOP, with a financial allocation of over 5.4 billion CZK (207.7 million EUR), and 28 IUDPs were implemented in 23 cities in the ROPs, with a total financial allocation of 12.7 billion CZK (488.5 million EUR) (Ministry for Regional Development of the Czech Republic 2014). The interventions in the IUDPs focussed on three main issues:

- improvement of the environment in problematic housing estates (usually revitalisation of concrete block of flats and their surroundings)—the IOP;
- improvement of deprived areas (usually usage of abandoned sites, improvement of accessibility to public services or traffic accessibility)—ROPs;
- development of territories with high growth potential (usually revitalisation of city centres and their surroundings)—ROPs.

Six general priorities were set for the IUDPs that should be used for the focus of integrated strategies: economic development, social integration, the environment, attractive cities, accessibility and mobility, and governance. All priorities aim to improve living conditions and economic development and thus have the potential to increase the cohesion of cities through the development of bridging and bonding social capital.

The IUDPs for interventions in ROPs were obligated to focus on at least three of the six priorities to achieve the synergistic effect of the interventions. However, the interventions from the IOP were clearly defined to improve the quality of housing, and the other priorities for the 'soft' investments were voluntary (Ministry for Regional Development of the Czech Republic 2012). Six cities within the main deprived areas were an exception. In these cases, socially excluded localities with significant Roma minorities were selected for pilot testing. They had an obligation to include social priorities for social inclusion financed from the Human Resources and Employment Operational Programme in addition to the priorities focussed on tangible investments for the revitalisation of public spaces and the regeneration of residential buildings.

Although the cities themselves could design the focus of priorities in the IUDPs, these integrated strategic plans mainly focussed on the priorities that allowed the realisation of investment projects based on the building of a basic infrastructure. These large investment projects, such as the insulation and revitalisation of old concrete block of flats or the construction of transport infrastructure (buying new mass transport means or terminals and stops), are not attractive for many nonprofit organisations to participate in. Some of the city representatives mention that setting the priorities included in the IUDPs was not conducive to involving the public and nonprofit organisations. Cooperation between municipalities and NPOs was closer in priorities focussed on the integration of socially excluded Roma communities, where the role of NPOs was to increase city cohesion more significantly.

The Ministry of Regional Development issued general guidelines for the implementation of integrated development approaches. Each participating city had to develop an IUDP strategic document of its own to identify and justify a suitable location for the implementation of the IUDP. The document should contain a list of suitable projects for implementation and explain the method of involvement of other relevant partners in the process of preparing and implementing the IUDP based on the partnership principle.

There were two stages of IUDP creation and implementation. At the first stage, the IUDPs (strategic documents) had to be prepared with an indicative list of expected project proposals. Moreover, the cites had to submit them to the specific calls for proposals in the IOP or ROPs, where their quality was assessed. At the second stage, individual project proposals were submitted to standard calls for proposals and had to successfully compete with other submitted projects. However, if they were a part of any IUDP, they received bonus points in the assessment (10%).

The main advantage of the integrated strategies should be a complex approach to solving problems in a certain area within the city that also promote cohesion in the city. In combination with sufficient financial sources allocated to the programmes, the strategies enabled quick implementation of many investments and non-investment projects concentrated in one area. Thus, the IUDPs represented a great opportunity for the development of Czech cities.

However, some representatives of Czech cities complained that this initial promising concept of cross-programme funding had practically disappeared through the practical implementation of the IUDP instruments and that easier and more accessible funding for diverse activities from multiple sources was not being disbursed. Based on the analysis of the IUDP documents, the funds were often used only for one or two priority axes, and all investments were financed only by the single operational programme.

We have found only minor differences in the application of the partnership principle with NPOs and civil society in particular cities. The cities often used similar channels to communicate with the general public, such as local municipal journals or magazines, official city web pages or public discussion meetings. It was common for the cooperation to mainly focus on informing the public about already established priorities in the IUDPs, rather than discussing the preparation of strategic documents from the very beginning. The city representatives also often used communication channels to familiarise the public with investment projects that were already prepared. At its preparation stages, the process was characterised as top-down implementation rather than the bottom-up generation of ideas based on multi-actor discussions. Thus, the partnership principle was fulfilled only formally as a necessary condition for obtaining funds.

In many instances, however, not only do cities have to lead an intensive discussion with the associates of the owners of residential units or housing cooperatives, which are not only the key beneficiaries, but they also have to prepare particular project applications for the grants. Oftentimes, they have an informal character or can take the form of NPOs, but they are not considered representatives of NPOs or civil society as such. However, these meetings can contribute to the increase of social capital and the cohesion of property owners within the community because cooperation between neighbours and communities was necessary to succeed in the implementation of particular projects.

More space was provided to representatives of civil society and NPOs in the Steering Committees of the IUDPs or in individual thematic working groups, which had a greater influence on the final list of projects implemented under the IUDPs. However, in some cases, representatives of the nonprofit sector were invited to the working groups with some delay (not before the implementation). City representatives admit that even they did not know about the activities of some nonprofit organisations operating in the city territory; they learned about their existence only when discussing investment plans or individual projects.

Undoubtedly, greater involvement of NPOs (but not citizens) was realised during the implementation of the particular projects under the IUDPs where NPOs were one of the eligible applicants. If NPOs brought a new project proposal that fit the priorities of approved IUDPs, it was possible to add it to the list of project proposals supported by the IUDPs. However, their projects have to be approved by the Steering Committee. In many cases, this was ensured by submission of the projects in partnership with the city.

Notably, in many cases, the IUDPs were one of the first approaches that required close cooperation and communication with the wider public and other partners at the local level, which was uncommon in all cities at that time. Just starting the process of participating with the public and discussing further development of cities or their specific parts was seen by some city representatives as one of the main benefits of the implementation of the IUDPs. Thus, the identification of partners and the initiation of discussions among them can be considered one of the main contributions of the IUDPs to the internal cohesion of Czech cities.

Moreover, city representatives add that cooperation between different entities is more intense with the new integrated tools used in the current programming period 2014–2020, where the principles of integrated development transferred into new integrated tools for the development of larger metropolitan areas are being implemented. There are two tools being implemented: ITIs for metropolitan areas with urban agglomerations of over 300 thousand inhabitants (7 territories) and Integrated Territorial Development Plans used for other key metropolitan areas with urban agglomerations of less than 300 thousand inhabitants (6 territories). In addition to these regional policy tools, there are also CLLDs, designed for Local Action Units, which have a dense network in the Czech Republic, but they only cover rural areas. It seems that some deficiencies related to multi-source financing have been removed and a broader focus on new integrated tools provides more room for other partners such as NPOs and civil society to be involved. Of course, the experience gained through the implementation of IUDPs in previous years also plays a role.

2.5 Discussion and Conclusions

The role of civic engagement and NPOs has attracted attention in recent research. Our study attempts to shed light on the role of these factors in territorial and social cohesion in cities. We concentrate on the implementation of the IUDPs funded by the EU Structural and Investment Funds in the Czech Republic.

The lessons learned from the implementation of integrated approaches to urban development in Czech cities show us that NPOs and civic engagement play an important though indirect role in promoting better living conditions and, consequently, social territorial cohesion. The implementation of the IUDPs was basically a technocratic issue. This is evidenced by the fact that representatives of NPOs were invited to participate in the creation and implementation of these plans but only actually helped formally approve the investment plans already prepared. However, this does not diminish the importance of improving public spaces and public infrastructure provided by this EU funding and its importance for the well-being of urban dwellers.

On the one hand, the IUDPs allowed cities to gain relatively large funds for investments that would otherwise take a very long time. Usually, the investment projects had already been prepared (with or without the participation of the public), and the justification of the IUDPs, including strategy and SWOT analysis, was prepared according to the investment plans. The IUDPs are useful tools that provided an approach to investment in certain parts of the cities. Maintenance of tangible results is the responsibility of cities and their budgets; thus, there is no doubt that they are sustainable. In the case of intangible results, they depend more on trust and social capital in cities. Maintaining such results falls beyond the direct control of the cities, as it is based more or less on the voluntary activities of people at both individual and organised levels.

On the other hand, the IUDPs were often not integrating complex investments. The cities usually decided on one or two main investment priorities from the IOP or the ROPs (especially reparation of concrete blocks of flats, road reconstruction, reconstruction of parks, etc.). Combination with other programmes was the exception. The cities rarely implemented soft investment projects, even if they were also funded. It was usually the role of NPOs to fill this gap and to contribute to social and territorial cohesion.

Three issues are important for NPOs and civic engagement in Czech cities and their goal of achieving social and territorial cohesion. First, low trust and social capital make it difficult for NPOs to take an active part in political decision-making processes. This concerns the issue of responsibility for political decisions and taking political responsibility for those decisions, which the current political system does not allow. A strong collaborative spirit, which is vital for place-based development approaches, is missing (Horlings et al. 2018). Thus, people and NPOs can express their needs, but it is up to local politicians to decide what priorities will be funded. Second, the current rise of local movements in Czech politics indicates conflicting interests in local governance. These conflicting interests can lead to the creation of coalitions that exclude social actors from the process (Trudelle et al. 2015) and diminish social and territorial cohesion processes. The dichotomous approach of political elites to civil society (Potůček 1999) and the low acceptance of NPOs as partners (Potluka et al. 2017) cause visible tensions. When the political representation does not help foster territorial and social cohesion, people start politically organising and set up their own movements, which provide them the social cohesion they need for political action. Third, bottom-up movements increase the relevance of local policies and slowly change the relationships in local policymaking (Bowden and Liddle 2018), increasing the feeling of belonging to the local population.

In summary, none of the political and investment tools can build social and territorial cohesion directly. It can only be done indirectly, step by step, by the long-term building of trust and the feeling of belonging to the community and the city. When neither private companies nor the public sector provides certain services, local NPOs can help and start to co-create them while serving as a bridge between people's needs and the provision of local public services. In fact, social and territorial cohesion belong among these issues, as NPOs are set up from the bottom-up. To achieve social and territorial cohesion, partnership among all three sectors is needed.

Still, an unanswered question remains—will local bottom-up movements be caught in the same trap as political parties and become disconnected from local needs? The intensified political work of movements can lead to their professionalisation and disconnection from local communities. Although it is a very interesting question, we leave it unanswered, as it falls beyond the scope of our current research.

References

Andreassen TA, Breit E, Legard S (2014) The making of 'professional amateurs'. Acta Sociol 57(4):325–340. https://doi.org/10.1177/0001699314552736

Andreotti A, Mingione E, Polizzi E (2012) Local welfare systems: a challenge for social cohesion. Urban Stud 49(9):1925–1940. https://doi.org/10.1177/0042098012444884

Baun M, Marek D (eds) (2008) EU cohesion policy after enlargement. Palgrave Macmillan, New York

Ben-Ner A, van Hoomissen T (1992) An empirical investigation of the joint determination of the size of the for-profit, nonprofit and government sectors. Ann Publ Coop Econ 63(3):391–415. https://doi.org/10.1111/j.1467-8292.1992.tb02098.x

Bowden A, Liddle J (2018) Evolving public sector roles in the leadership of place-based partnerships: from controlling to influencing policy? Regional Studies 52:1–11. https://doi.org/10.1080/00343404.2017.1346369

Brandsen T, Pestoff V (2006) Co-production, the third sector and the delivery of public services. Publ Manag Rev 8(4):493–501. https://doi.org/10.1080/14719030601022874

Capello R (2018) Cohesion policies and the creation of a European identity: the role of territorial identity. JCMS: J Common Market Stud 56(3):489–503. https://doi.org/10.1111/jcms.12611

Cassiers T, Kesteloot C (2012) Socio-spatial inequalities and social cohesion in European cities. Urban Stud 49(9):1909–1924. https://doi.org/10.1177/0042098012444888

Colini L, Tripodi L (2010). Sustainable urban development: implementation praxis of Art 8 (n°2009.CE.16.0.AT.109). Brussels, Belgium: European Commission, Directorate-general for Regional Policy, Unit Urban development, territorial cohesion

Cornforth C, Hayes JP, Vangen S (2015) Nonprofit-public collaborations: understanding governance dynamics. Nonprofit Voluntary Sect Q 44(4):775–795. https://doi.org/10.1177/0899764014532836

Dąbrowski M, Bachtler J, Bafoil F (2014) Challenges of multi-level governance and partnership: drawing lessons from European Union cohesion policy. Eur Urban Reg Stud 21(4):355–363. https://doi.org/10.1177/0969776414533020

EC (2004) Project Cycle Management Guidelines. Brussels, EC

EC (2013) Flash Eurobarometer 373: Europeans' engagement in participatory democracy

EC (2014a) Community-led local development—cohesion policy 2014–2020, Factsheet. Brussels, Belgium, European Commission

EC (2014b) The European code of conduct on partnership in the framework of the European Structural and Investment Funds. Brussels, EC

EC (2014c) Integrated territorial investment—cohesion policy 2014–2020: Factsheet. Brussels, Belgium, European Commission

Eizaguirre S, Pradel M, Terrones A, Martinez-Celorrio X, Garcia M (2012) Multilevel governance and social cohesion: bringing back conflict in citizenship practices. Urban Stud 49(9):1999–2016. https://doi.org/10.1177/0042098012444890

ESS Round 7: European Social Survey Round 7 Data (2014) Data file edition 2.1. NSD—Norwegian Centre for Research Data, Norway—Data Archive and distributor of ESS data for ESS ERIC. Retrieved from http://www.europeansocialsurvey.org/data/download.html?r=7

EU (2006) Regulation (ec) no 1080/2006 of the European Parliament and of the council of 5 July 2006 on the European Regional Development Fund and repealing Regulation (EC) No 1783/1999. Official Journal of the European Union

Frič P (2004) Politický vývoj ve střední Evropě po r. 1989 a jeho dopad na neziskový sektor. Retrieved 15 May 2015 from http://www.ceses.cuni.cz/CESES-20-version1-sesit04_03_fric.pdf

Hansmann Hb (1980) The role of nonprofit enterprise. Yale Law J 89(5):835–903. https://doi.org/10.2307/796089

Hanssen Gs (2010) Ensuring local community interests in market-oriented urban planning? The role of local politicians. Environ Plann C: Govern Policy 28(4):714–732. https://doi.org/10.1068/c09102

Horlings LG, Roep D, Wellbrock W (2018) The role of leadership in place-based development and building institutional arrangements. Local Econ: J Local Econ Policy Unit, 026909421876305. https://doi.org/10.1177/0269094218763050
Howard MM (2011) Civil society in post-communist Europe. In: Edwards M (ed) The Oxford handbook of civil society. Oxford University Press, New York
Karp JA, Milazzo C (2015) Democratic scepticism and political participation in Europe. J Elections, Public Opin Part 25(1):97–110. https://doi.org/10.1080/17457289.2014.996157
Keil R (2011) The global city comes home: internalised globalisation in Frankfurt Rhine-Main. Urban Stud 48(12):2495–2517. https://doi.org/10.1177/0042098011411946
Lavasseur K (2018) Co-producing accountability? Drawing conclusions from non-profit child care services in Manitoba. Can Publ Adm 61(1):26–44
Luria G, Cnaan Ra, Boehm A (2015) National culture and prosocial behaviors: Results from 66 countries. Nonprofit Voluntary Sect Q 44(5):1041–1065. https://doi.org/10.1177/0899764014554456
Mascherini M, Vidoni D, Manca AR (2011) Exploring the determinants of civil participation in 14 European countries: one-size-fits none. Eur Sociol Rev 27(6):790–807
Ministry for Regional Development of the Czech Republic (2012) Metodický pokyn Ministerstva pro místní rozvoj k hlavním zásadám pro přípravu, hodnocení a schvalování Integrovaného plánu rozvoje města. Ministry for Regional Development of the Czech Republic, Prague. Retrieved from https://mmr.cz/getmedia/7f47c789-e4e1-4d16-9f25-8cad4f8b8b7d/Metodicky-pokyn-MMR-k-hlavnim-zasadam-pro-pripravu
Ministry for Regional Development of the Czech Republic (2014) *Výsledky intervencí ze strukturálních fondů a Fondu soudržnosti dosažené v tematických oblastech v letech 2007–2013*. Prague: Ministry for Regional Development of the Czech Republic Retrieved from http://dotaceeu.cz/getmedia/e519cda8-7280-40b8-b5d1-451a82f555ff/Vysledky-intervenci_verze_s_ISBN.pdf?width=0&height=0
Nałęcz S, Leś E, Pieliński B (2015) Poland: a new model of government–nonprofit relations for the east? VOLUNTAS: Int J Voluntary Nonprofit Organ 26(6):2351–2378. https://doi.org/10.1007/s11266-015-9653-4
Novy A, Swiatek DC, Moulaert F (2012) Social cohesion: a conceptual and political elucidation. Urban Stud 49(9):1873–1889. https://doi.org/10.1177/0042098012444878
Paarlberg LE, Yoshioka T (2016) The impact of local economic structure on community philanthropy. Public Adm Rev 76(2):340–350. https://doi.org/10.1111/puar.12442
Pestoff V (2012) Co-production and third sector social services in Europe: some concepts and evidence. VOLUNTAS: Int J Voluntary Nonprofit Organ 23(4):1102–1118. https://doi.org/10.1007/s11266-012-9308-7
Porter M (2008) On competition. Harvard Business School Publishing, Boston
Potluka O, Perez M (2018, forthcoming) Do candidates from non-profit organisations who adopt party political values improve their chances of electoral success? Policy and Politics
Potluka O, Idczak P, Kalman J, Kostal C, Musialkowska I, Ovádek M, Tholtová B (2014) Local Participation in Visegrad Countries: How Are Politicians Rooted in the Civil Society?. IREAS, Institut pro strukturální politiku, Praha
Potluka O, Špaček M, Remr J (2017) NGOs as partners: obstacles in the EU cohesion policy? Ekonomický časopis 65(8):715–736
Potluka O, Kalman J, Musiałkowska I, Idczak P (2018, forthcoming) Civic engagement in local politics in Central Europe. In Kerley R, Dunning PT, Liddle J (eds) The Longman handbook on international local government. Abingdon, Routledge
Potůček M (1999) Havel versus Klaus: public policy making in the Czech Republic. Journal of Comparative Policy Analysis 1(2):163–176. https://doi.org/10.1023/a:1010023002265
Putnam RD (2000) Bowling alone—the collapse and revival of American community. Simon & Schuster, New York
Rose-Ackerman S (2007) From elections to democracy in Central Europe: public participation and the role of civil society. East Eur Polit Soc 21(1):31–47. https://doi.org/10.1177/0888325406297132

Škorňa D (2011) The usage of integrated approaches in development of territories. Regionální studia 2:58–70
Strom E (2008) Rethinking politics of the downtown development. J Urban Aff 30(1):37–61. https://doi.org/10.1111/j.1467-9906.2007.00373.x
Su C (2018) Managed participation: city agencies and micropolitics in participatory budgeting. Nonprofit Voluntary Sect Q, 089976401875702. https://doi.org/10.1177/0899764018757029
Szokolai Z (2012) Integrated territorial development and their application to ETC in the cohesion policy 2014–2020. http://www.eukn.eu/fileadmin/Lib/files/EUKN/2013/Presentation_Integrated_Instruments_presentation_Zsolt_SZOKOLAI_Commission_.pdf
Trudelle C, Klein J-L, Fontan J-M, Tremblay D-G (2015) Urban conflicts and socio-territorial cohesion consensus building and compromise in the Saint-Michel neighbourhood in Montreal. Can J Urban Res 24(2):138–157
UN (2015) World urbanization prospects: the 2014 revision. Retrieved 17 Nov 2015 from http://esa.un.org/unpd/wup/FinalReport/WUP2014-Report.pdf
Vallaster C, von Wallpach S (2018) Brand strategy co-creation in a nonprofit context: a strategy-as-practice approach. Nonprofit Voluntary Sect Q, 089976401877637. https://doi.org/10.1177/0899764018776373
Vamstad J (2012) Co-production and service quality: the case of cooperative childcare in Sweden. VOLUNTAS: Int J Voluntary Nonprofit Organ 23(4):1173–1188. https://doi.org/10.1007/s11266-012-9312-y
van der Zwet A, Bachtler J, Ferry M, McMaster I, Miller S (2017) Integrated territorial and urban strategies: how are ESIF adding value in 2014–2020?. Eur Comm, Direct—Gen Reg Urban Policy, Brussels, Belgium
Vidoni D, Mascherini M, Manca AR (2009) Scouting for drivers of the European knowledge society: the role of social participation, Berlin, Heidelberg
Voorberg WH, Bekkers VJJM, Tummers LG (2015) A systematic review of co-creation and co-production: embarking on the social innovation journey. Public Manag Rev 17(9):1333–1357. https://doi.org/10.1080/14719037.2014.930505
Weisbrod BA (1975) Toward a theory of the voluntary non-profit sector in a three-sector economy. In: Phelps ES (ed) Altruism, morality, and economic theory. Russell Sage Foundation, New York

Oto Potluka is a Senior Researcher at the Centre for Philanthropy at the University of Basel, Switzerland. His main concerns are evaluations of the impact of public expenditure programmes. He specialises on the EU Cohesion policy and regional development, including the role of the civil society in development policies. He is a member of Regional Studies Association, European Evaluation Society, Swiss Evaluation Society, and Czech Evaluation Society.

Martin Špaček is a Ph.D. candidate at Comenius University in Bratislava. His work concerns mainly international research activities with the focus on regional polarisation in Central and Eastern Europe in relation to cross-border cooperation and social innovations in marginalised rural areas. He is a member of Czech Evaluation Society as a practical evaluator of several public expenditure programmes. He previously participated as an Early Stage Researcher in the Innovative Training Network RegPol2 on 'Socio-Economic and Political Responses to Regional Polarization in Central and Eastern Europe' at SPECTRA CE, Slovak University of Technology in Bratislava.

Chapter 3
Territorial Cohesion: The Economy and Welfare of Cities

Jacek Zaucha and Tomasz Komornicki

Abstract The paper examines how cities contribute to three dimensions of territorial cohesion: polycentricity, territorial governance and the socio-economic growth. The focus is on the role of cities and Functional Urban Areas (FUAs) as producers of well-being and prosperity in space. This issue is examined on the basis of the recent economic models. However, the economic approach is filtered through the lenses of the territorial cohesion, i.e. impact of growth on polycentricity, and territorial governance (place-based policy design). The theoretical considerations are illustrated by empirical findings mainly from Poland and Central Europe illustrating interplay between growth, polycentricity and territorial governance. On that basis the authors propose how to better integrate cities and FUAs in pursuing territorial cohesion at regional, national, transborder and EU level. The paper provides some plausible answers how to make use of urban economic growth for reinforcing key aspects of territorial cohesion such as vertical and horizontal co-operation, connectivity, urban–rural interactions and securing access to services of general economic interest.

Keywords Welfare of cities · Territorial cohesion · Polycentrism
Territorial governance · Territorial cooperation

3.1 Introduction

The chapter examines the ways in which cities and their surrounding urban regions contribute to the goal of territorial cohesion. Its focus is the role of cities and

J. Zaucha (✉)
Faculty of Economics, University of Gdańsk and Institute for Development, Gdańsk and Sopot, Poland
e-mail: jacek.zaucha@gmail.com

T. Komornicki
Department of Spatial Organization, Institute of Geography and Spatial Organization Polish Academy of Sciences, Warsaw, Poland
e-mail: t.komorn@twarda.pan.pl

E. Medeiros (ed.), *Territorial Cohesion*, The Urban Book Series,
https://doi.org/10.1007/978-3-030-03386-6_3

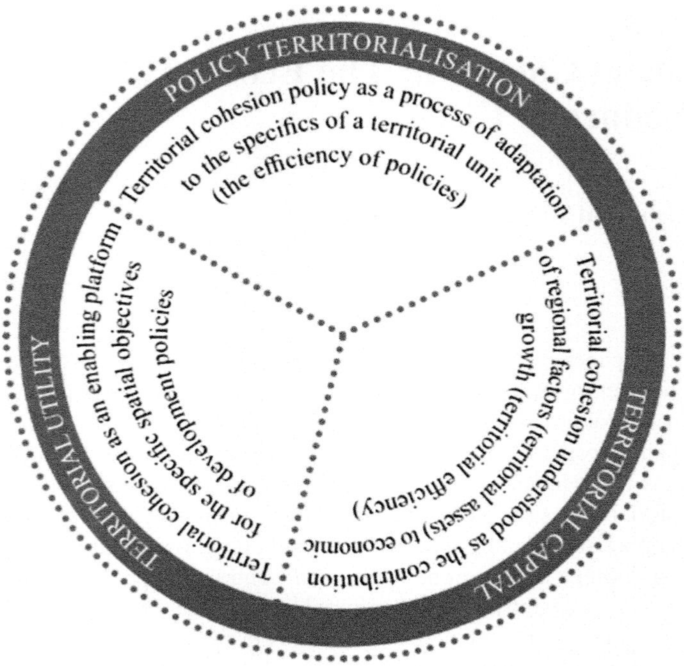

Fig. 3.1 Key dimensions of territorial cohesion. *Source* Zaucha and Szlachta (2017, p. 47)

Functional Urban Areas (FUAs) as producers of well-being and prosperity in terms of space. In this chapter, the definition of territorial cohesion is based on two models developed by the Polish Institute of Development and Eduardo Medeiros (Star model), respectively. Polish researchers (Zaucha and Szlachta 2017, 46–47) identified three dimensions of territorial cohesion: policy territorialisation, territorial utility and territorial capital (Fig. 3.1).

As a result of the breadth of the aforementioned three dimensions, this chapter will focus only on the examination of their specific aspects in the urban context. These aspects have been selected in accordance with the Medeiros (2014, 20) Star model which considers polycentrism, economic and social cohesion, co-operation/governance and environmental sustainability as key dimensions of territorial cohesion. Therefore, in this chapter, territorial utility (i.e. trade-offs between spatial and economic goals) will be analysed through the lens of the concept of polycentrism, policy territorialisation (i.e. adapting policies to the specificity of territorial units—see Doucet et al. 2014) will be interpreted as a part of the multilevel governance paradigm and the discussion on territorial capital will be limited to economic cohesion and growth as well as the role of cities in this process.

The scope and length of this chapter do not allow for a more in-depth analysis of the origin, essence and conceptualization of the notion of territorial cohesion. Two comprehensive overviews on that issue have been published recently (Medeiros 2016; Zaucha and Szlachta 2017). Moreover, Faludi (2016) has put forward an inspiring proposal on the future of territorial cohesion interpreted as the conceptualising and re-conceptualising of territorial relations which constitute territories in their functional dimension. While these considerations will stand at the background of this chapter, the key aim here will be the close analysis of the contribution of the cities to the territorial cohesion rather than a debate concerning its nature.

European countries provide different definitions for both urban and rural areas. As a result, many statistical data is incomparable. Suburbanisation processes add to the difficulty in interpreting statistical data. Many cities are losing their population and also, at times, the economic potential located within their administrative boundaries. Simultaneously, the Functional Urban Area (FUA) of the same urban centres is growing. Therefore, the correct recognition of these processes is of key importance in assessing the development potential of cities. This is the reason why the 'functional regions' themselves represent one of the territorial keys defined in the Polish Presidency Report (cf. Zaucha et al. 2014). The tools of spatial econometrics are also helpful in overcoming the interpretation bias (Ciołek 2017; Ciolek and Brodzicki 2015, 2016, 2017b).

3.2 Cities as Engines of Growth and Prosperity—Territorial Capital

It is a statistical fact that the lion's share of the world GDP is produced in urban areas. According to the World Bank, 80% of the global GDP is generated in cities, leading to the conclusion that urbanisation can contribute to growth (see also Yusuf and Leipziger 2014). This issue has been studied empirically in many research projects and the general conclusion is that growth and prosperity go hand in hand with spatial concentration of production (World Bank 2009). Developed economies have experienced this correlation during the period of their industrialization, while nowadays it is obvious in emerging markets like China. For instance, Duranton (2014) estimated that 1% point of urbanisation is associated with an approximately 5%-point increase of the GDP per capita, and that 60% of the variance of the prosperity level across countries could be explained by this single variable. Du (2017) has confirmed the causal relationship between land urbanisation quality and economic growth in China. According to Mukhopadhyay (2018), the per capita net value added in India's urban areas in 2011–2012 was 2.5 times higher than in the rural ones.

Certainly, detailed analysis reveals more complex and non-uniform trends. The Glaeser's (2013, 7) estimates of the slope of coefficient for the log of GDP per capita regressed on urbanisation are much lower for developing countries than for the rest of the world. Arouri et al. (2014) revealed an inverted U-shape relationship between the urban population share and the per capita GDP in Africa with the threshold value of urbanisation at approximately 73%. After exceeding this point, higher urbanisation goes hand in hand with the decrease of the GDP per capita. This observation confirms Henderson's (2003) well-known findings. Similarly, Brülhart and Sbergami (2009) estimated that agglomeration enhances GDP growth only up to a certain level of prosperity roughly equalling USD 10,000 of GDP per capita.

Arouri et al. (2014, 14) have also pinpointed an important role of human capital as a mediating variable between growth and urbanisation. They concluded that 'a good level of human development benefits the growth effect of urbanization, while a weak human capital development amplifies negative effects of urbanization': Mukhopadhyay (2018) underlines that during the rapid GDP growth (above 8% per annum) period in India (May 2004 to December 2011), the rate of urbanisation increased marginally by 1.2% point. Urbanisation also results in some undesirable phenomena negatively affecting growth (this explains the aforementioned U-shape relationship between GDP growth and urbanisation). For instance, Colenbrander (2016) has pointed out that one in seven people in the world lives in poverty in urban areas and that such a situation substantially reduces the ability of cities to deliver growth. The author calls for poverty reduction, necessary in allowing cities to achieve their full growth potential in the long run.

Moreover, the relationship between urbanisation and growth plays differently for various types of cities. According to the World Bank (2015), competitive cities drive a disproportionate job growth and increased income and productivity. According to World Bank's estimates, only 72% of cities outperformed their countries in terms of GDP growth and can be classified as competitive ones. Mills and McDonald (2017, xiii) have noticed a large variation in terms of economic performance among US metropolitan areas. Mukhopadhyay (2018) highlights the in situ pattern of urbanisation in India and draws attention to the fact that smaller cities attract roughly 40% of migrants and are important growth engines in this country. In Poland, the situation is different. According to the World Bank (2017), growth is concentrated in large urban regions, whereas regions without such growth engines suffer from a relative economic slowdown.

Regardless of these differences, it is obvious that cities with their surroundings (urban areas) and networking between socio-economic agents located there play a pivotal role for growth and prosperity. The city-induced processes either enhance or slow down the performance of economic agents. Therefore, cities themselves belong to the so-called territorial capital, i.e. an unmovable or territorially bound spatial asset that cannot be easily replicated elsewhere. As pointed out by Markowski (2017, 70) 'territorial capital is inseparably related with the urban system'. Cities and urban regions were considered dimensions of territorial capital in Poland (Komornicki and Ciołek 2017) although, bearing in mind the original operationalisation of this notion by Camagni (2008), one should rather see urban areas as a spatial frame for the interplay between various elements of the territorial capital such as numerous types of externalities, relational capital, human capital, collective goods, public goods and many others.

According to mainstream economics, cities offer agglomeration economies, i.e. specific externalities stemming from the clustering of economic activities in the same place. This observation is usually attributed to Alfred Marshall (1879, 1890) who first described the mechanism of enhanced productivity as a result of clustering processes. This phenomenon is broadly described in the relevant literature (e.g. Ellison et al. 2010; Zaucha 2007) and does not require further discussion in this chapter. New economic geography (NEG) has extended this understanding of the cities' economic role by describing them as an outcome of the interplay between centripetal and centrifugal economic forces. The first one can be attributed to economies of agglomeration in the situation of monopolistic competition and the clear preference for variety among consumers. The second one is mainly caused by distance, i.e. the costs of serving markets outside the given urban region.

With diminished costs of trade and exchange (in terms of transport but also cultural barriers), economies of agglomeration prevail and the spatial pattern of production tends to be more concentrated. Usually, larger cities gain and smaller cities located outside the shadow of large cities and rural areas lose in terms of economic activities and, consequently, also population. This trend can be hypothetically reversed when the trade costs will become close to null and when territorial capital located outside the urban areas (e.g. living in a surrounding area close to nature) can outweigh the agglomeration benefits and can induce a reallocation of economic activities or people outside the urban areas since some employees can work from the distance (Zaucha et al. 2014, 249). The contemporary NEG models (new–new economic geography) assume some heterogeneity, not only of tastes but also of companies according to the Melitz-type model of monopolistic competition. Such heterogeneity acts as an additional centrifugal force (Ottaviano 2010).

According to Baldwin and Okubo (2009), only the most productive companies cluster in large urban regions whereas the less productive ones move to the periphery. Adjusting for differences in the capitals, Forslid and Okubo (2010) have provided evidence that clustering in urban regions concerns companies with high capital intensity and high productivity as well as enterprises with low capital intensity and capital productivity. More in-depth analyses of the spatial effects of economic activities in line with the NEG models are available in the relevant literature (see, for instance, Brodzicki 2018) and can be interpreted as evidence for the important economic role of cities and urban areas at a macro, mezzo and even micro-economic level.

Cities/urban regions gather a high concentration of human and physical capital, whilst providing the home market effect, and facilitating knowledge production. According to recent analysis, the spillovers from the 'urban economic hills' are spatially limited and extend to 175–200 km in the case of total factor productivity (Ciołek Brodzicki 2017a), 300 km in terms of knowledge (Bottazzi and Peri 2003), and 250–500 km for technology diffusion (Moreno et al. 2005). As a result of the above, cities/urban regions shape patterns of spatial development.

NEG attaches great importance to distance as a pivotal force in shaping spatial development. The development opportunities of cities are conditioned, among other factors, by their location in the pattern of other urban units, in particular other large urban centres. The measure that reveals this position is the potential accessibility index which also takes into account the city's size (either population or GDP), distance from other centres (time or cost) and the assumed distance-decay function. Such accessibility can enrich an analysis that is based on the traditional indicators of socio-economic development (cf. Spiekermann et al. 2013; Rosik et al. 2015). In short-term analyses, however, it is difficult to indicate the existence of dependence between accessibility and growth.

The role of accessibility as a factor in economic growth is often questioned. At the same time, studies corroborating this particular proposition are frequently conducted in countries characterised by a dense road and/or rail network. Numerous other studies point out that, while the development of infrastructure is not a sufficient condition for economic development, it is undoubtedly a necessary precondition for activating the growth processes. When perceived from a macroeconomic global perspective, the impact of infrastructure development on the economy of bigger cities is of an undisputed importance (Komornicki 2013). The few investigations concerning the problem that were carried out in the context of Central-Eastern Europe include the report by Cieślik and Rokicki (2013) which refers to the layout of the Polish provinces and metropolitan areas. These authors demonstrate that there exists a statistical interdependence between the development of the metropolises (their economic potential) and the state of road development and accessibility. Such correlation is also confirmed by the research done within the GRINCOH project (Komornicki et al. 2014).

However, such a conceptualization of the role of cities and urban regions in the territorial capital context as proposed by NEG (focusing on distance and economies of agglomeration) seems insufficient. It is equally important to consider a number of ideas pertaining to heterodox economics in particular institutions (Zaucha 2014). Cities host important economic institutions and themselves represent specific territorial institutions—as such, they sustain an idea covered by the concept of local milieu. Despite the frequently quoted proposal by North (1990, 3) who defined institutions as 'the rules of the game in a society or, more formally', as 'the humanly devised constraints that shape human interaction', institutions lack a uniformly accepted definition in literature. However, a broad consensus exists that they comprise both hard and soft types of institutions. The first type can be characterised by phenomena like 'constitutions, laws, charters, bylaws and regulations, as well as elements such as the rule of law and property rights and contract and competition' (Rodríguez-Pose 2013, 10). Soft or informal institutions cover 'norms, traditions and social conventions, interpersonal contacts, relationships, and informal networks' (Rodríguez-Pose 2013, 10) and create a core of social capital as well as contribute to trust building.

Some scholars (e.g. Staniek 2007) also consider organisations as components of an institutional tissue which includes local or regional governments. Institutions create a frame of economic governance described by Dixit (2009, 5) as 'the structure and functioning of the legal and social institutions which supports economic activity and economic transactions by protecting property rights, enforcing contracts, and taking collective action to provide a physical and organizational infrastructure'. The recent research by Rodríguez-Pose and Ketterer (2016) has evidenced that the lack of improvement in the quality of institutions limits growth, particularly in low-growth regions. Resultantly, the 7th Cohesion report (EC 2017, 136) puts forward the opinion that the 'quality of government matters for social and economic development across the EU and that it is an important determinant of regional growth'. The same report also includes the following strong conviction: 'improvements in the quality of institutions appear to have been among the most consistent factors underlying economic growth and resilience across the EU' (EC 2017, 141).

The institutional performance of cities and urban regions, understood as a nexus of various social and administrative interactions, is therefore of key importance for regional and national growth and prosperity. Cities and urban regions are distinguished from other types of territories by the intensity of such networks, relations and negotiation patterns. They can both boost development as factories of social innovations, local and regional leadership, trust and policy transparency or they can lower stimuli in the case of the malfunctioning of cities and urban regions. However, when compared to pure economic interactions, the institutional aspects present in cities and urban regions add an important governance dimension, i.e. ability to learn from previous failures, predict the future and even influence the future course of actions and be prepared to cope with future problems. Cities and/or urban regions might create, therefore, a so-called adaptive capacity which is a key mechanism in securing evolutionary resilience understood as the 'ability of complex socioecological systems to change, adapt, or transform in response to stresses and strains' (Davoudi et al. 2016, 713).

3.3 The Case of Polish Cities as Growth Engines

In Poland's case, the distinction between urban and rural areas has deep historical roots and is formally made on the basis of administrative criteria. Definite administrative units (municipalities) are granted 'urban rights' by a motion from the Council of Ministers. There are, formally, 1571 rural and 602 urban–rural municipalities in Poland. Altogether, the formally rural areas encompass 93% of the country's area and are inhabited by approximately 39% of the total population (as of 2015; Komornicki 2016). According to data from the GUS, the urbanisation index has ceased to go up since 1998 (Bański 2008), a fact which is sometimes interpreted as a halting of the rural–urban migrations. This interpretation is incorrect. Since the beginning of the 1990s, the population of the biggest cities has been decreasing while the population of the surrounding rural areas has been growing (Komornicki 2016). As pointed out by Markowski (2017, 70), based on official data, the share of inhabitants residing in cities amounts to 61%, whereas the actual share of residents of the urbanised areas reaches 80% in Poland.

Therefore, in order to correctly assess the potential of larger centres, it is very important to properly delimit functional areas, especially metropolitan areas. In Poland, delimitations were initially carried out based on a set of supralocal features (e.g. delimitation by Smętkowski et al. 2008). However, this was not an optimal approach. A delimitation of areas based on practical functions, in particular commuting to work (Śleszyński 2013) and to services of general economic interest, should be considered a more adequate and promising solution.

Śleszyński (2013) defines the FUA as a spatially cohesive zone of the city's impact. The result of the delimitation in FUA of Polish regional capitals (18 largest

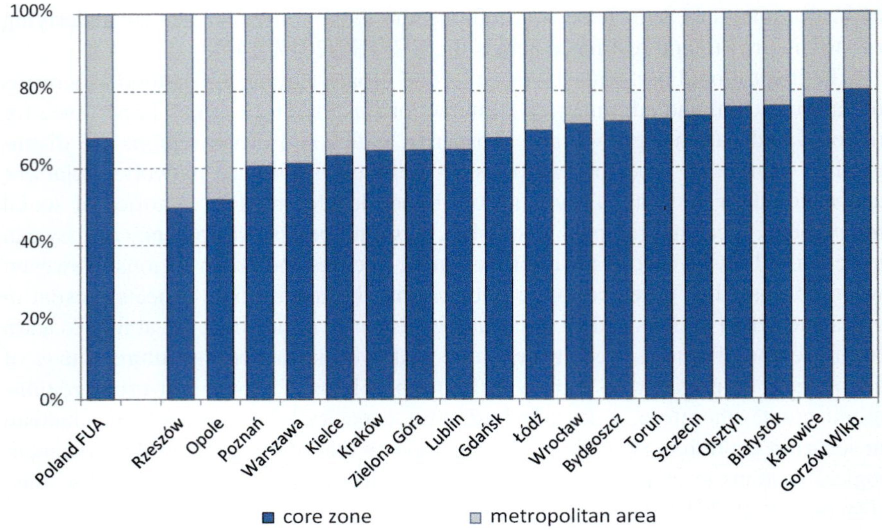

Fig. 3.2 Structure of the residents of the Functional Urban Areas of Polish regional capitals by share of inhabitants in the core and the external areas. *Source* Komornicki and Czapiewski (2016)

cities) is a list of 299 communes covered by those FUAs, including 33 core ones and 266 located in external zones. They occupy a total area of 32.6 thousand km^2 (of which core municipalities are 4.7 thousand km^2) and provide homes for 14.1 million inhabitants (9.6 million). The share of core and external areas in the FUA of regional capitals is very different (Fig. 3.2, Komornicki and Czapiewski 2016) as it depends on the progress of suburbanisation but also on several features of the local settlement structure (e.g. it also covers densely populated rural areas with strong multidirectional commuting as in the Polish Podkarpackie region).

The role of the metropolis has been growing throughout Poland's transformation period. This growth is connected to deindustrialization and the parallel development of the service sector concentrated in the largest polyfunctional cities. The polycentric character of the settlement network manifests itself primarily in the distribution of the population. With regard to the share of the largest cities in the production of GDP, the level of concentration is significantly higher. The European studies (ESPON 2004) indicate the existence of eight metropolises in Poland (MEGAs), including one MEGA of category 3 (Warsaw) and seven MEGAs of category 4, i.e. emerging (Gdańsk/Tri-city, Poznań, Wrocław, Szczecin, Katowice, Łódź, Kraków). While these cities and their FUAs offer shelter for about 36% of the population of Poland, they produce nearly 50% of GDP. In both cases, the concentration tends to increase, but in the case of the GDP, the increase is clearly faster. In the years 2000–2015, the share of eight MEGAs in generating GDP increased by 2.5% points. This value, however, is the result of the very different situations of individual metropolises (Table 3.1).

Table 3.1 The population and GDP of the Polish MEGAs* in 2000 and 2015 in absolute and relative terms

MEGAs	Population				GDP			
	2000		2015		2000		2015	
	Number	%	Number	%	mln PLN	%	mln PLN	%
Łódź	1,171,934	3.1	1,086,993	2.8	23,482	3.1	56,526	3.1
Warszawa	3,095,420	8.1	2,987,482	7.8	119,743	16.0	305,815	17.0
Kraków	1,402,529	3.7	1,476,837	3.8	30,496	4.1	83,102	4.6
Katowice	3,572,484	9.3	3,384,669	8.8	75,723	10.1	169,167	9.4
Poznań	1,095,000	2.9	1,176,355	3.1	34,022	4.6	86,283	4.8
Szczecin	944,809	2.5	918,159	2.4	20,523	2.7	41,697	2.3
Wrocław	1,162,154	3.0	1,217,562	3.2	26,170	3.5	76,456	4.2
Gdańsk	1,215,873	3.2	1,313,918	3.4	27,599	3.7	69,120	3.8
Total	13,660,203	35.7	13,561,975	35.3	357,758	47.9	888,166	49.4
POLAND	38,253,955	100.0	38,437,239	100.0	747,032	100.0	1,799,392	100.0
'Big Five'	7,970,976	20.8	8,172,154	21.3	238,030	31.9	620,776	34.5

* The eight Polish MEGAs as defined in the ESPON 1.1.1 project were adopted. The NUTS3 units corresponding to their cores and the adjacent areas (suburban area) were included

The five Polish metropolises ('Big Fives' the 'Big Five') account for a huge part of the dynamics of the Polish GDP: Warsaw, Poznań, Wrocław, Kraków and Gdańsk/Tri-city. In this group's case, the increase in the economic role reaches 2.5% points, with Warsaw demonstrating the highest dynamics (increase by 1% point) and Wrocław following close behind (0.7% points). The other MEGAs (Katowice, Łódź and Szczecin) are losing their leading positions in the Polish economy as a result of the collapse of a number of industries and migration outflow (Komornicki and Czapiewski 2016).

The analyses presented prove that achieving the critical mass of development depends on a number of factors. The number of people within an individual city is an imperfect and often misleading indicator. Therefore, for the sake of correct policy programming, it is necessary to include entire functional areas and a parallel assessment of the economic potential of a given FUA is fundamentally important. The position of the centre in the city network, determining the actual polycentrism of the system (see below) is also of great importance. Shortening the time distance to other centres and to the immediate hinterland may temporarily reduce the effects of certain cities' depopulation. In the accessibility research carried out for the exigencies of the Polish Ministry of Development (Komornicki et al. 2014), a conclusion was reached regarding the existence of a periodic compensation of the actual population loss by an improved access to regional labour markets. Such compensation is possible in the period of rapid infrastructure development, which, for example, took place in the Lodz region. The number of people living within the 60-min isochrone from the city centre has not decreased, despite the outflow of inhabitants. After completing the basic network of transport links, the dynamics of accessibility improvement decelerates and can no longer balance the natural and migration losses. This occurrence confirms that while transport is important at specific stages of urban development, it cannot be the only basis for the enhancement of the urban areas' performance.

3.4 The Urban Governance Dimension—Policy Territorialisation as a Result of the Existence of Cities and Urban Areas

In the previous section, cities and urban regions were presented as growth engines. They contribute to national, regional and local prosperity. The institutional aspects underline the importance of urban government in the fulfilment of this role. In general, cities can be treated as vehicles of policy territorialisation. Due to different allotments of development assets (including territorial capital) and due to the different institutional setups of the urban centres and their FUAs, the existence of cities undermines the efficiency of uniform policy solutions. This is true even for traditionally spatially uniform macroeconomic policies. For instance, Nijkamp puts forth arguments for the territorialisation of monetary policy (Nijkamp 2010, 17–18).

According to the OECD (1997, 143), policy territorialisation can be linked to the coordination of local actions by the administrative and functional levels. Zaucha et al. (2014, 249) describe the territorialisation of policy as introducing the territorial dimension into this kind of governance. 'In practice, this may mean that the policy is conducted in such a way as to take into account the territorial context (i.e. different objectives and different tools to achieve them for various territorial units or areas) and/or including territorial capital. Policy based on the principle of an integrated territorial approach (i.e. territorially sensitive policy) emphasises the endogenous potential, both as existing potential and as potential that could be achieved by the territory, and adapts the intervention to the spatial (or territorial) context of local or regional specifics' (Szlachta et al. 2017, 51).

A multilevel territorial governance seems to be one of the key vehicles of policy territorialisation from an urban perspective. According to the Tango ESPON (2013, 11) project, territorial governance refers to the formulation and implementation of public policies programmes and projects for the development of a place/territory by (i) coordinating actions of actors and institutions, (ii) integrating policy sectors, (iii) mobilising stakeholder participation, (iv) being adaptive to changing contexts and (v) realising place-based/territorial specificities and impacts. In short, such a governance focuses on territorial assets, integrative solutions and the creation of adaptive capacity for a given place.

In terms of governance, a key challenge lies in the limits of the welfare function of cities and urban regions. The local or regional government and local and regional networks usually put forward and prioritise the welfare or quality of life of local and regional population, whereas they attach much less attention to the fact that the performance of cities and urban regions is decisive for the welfare and quality of life at a national and even a supranational level. A problem exists in coping with the positive and negative externalities of city functioning in relation to their surrounding areas. Basically, there are two mechanisms of multilevel governance:

(a) Horizontal multilevel territorial governance between cities and other settlement centres in the adjacent areas;
(b) Vertical multilevel territorial governance encompassing districts, cities, regions and national level of policymaking.

Another complication comes from the fact that cities can be interpreted as fixed territorial units but also as territorial functional nexus constituted by various relations and interdependencies created or run or empowered by socio-economic entities that are somehow attached (in symbolic, cultural, economic or other terms) to a given city (or to cities as such). Faludi (2015) argues that there are plenty of no-man's lands with no clearly assigned formal governmental responsibilities. For a no-man's land—and this is typical case for FUAs—co-operation and voluntary coordination are among the most reasonable and effective governance mechanisms.

In terms of multilevel vertical governance, one of the biggest dilemmas regards the ways in which to utilise cities and urban regions as growth engines while avoiding an egoistic, rent-seeking behaviour and development at the expense of national or EU goals and values. As pointed out by Zaucha (2011) who reviewed the literature concerning this specific topic, 'municipal authorities are interested mainly in assuring high living standards for the citizens—their voters. Therefore, they often disregard the role of their cities in the development of larger territorial units'. Barca's (2009) place-based approach has provided one of the plausible answers regarding its mitigation in practice (see also Zaucha et al. 2013, 2017) since it requires a territorial dialogue. However, territorially blind approach has attracted the attention of the decision-makers in recent years at the most.

One can observe in Europe an inflexible inclination of policy concentration in terms of deciding on EU structural and investment funds. Poland remains among the few EU countries with a relatively high level of a decentralisation of command over these funds (up to regional level). In addition, the EU Urban Agenda has failed to strengthen place-based policy paradigms in the urban context. It concentrates more on solving structural city problems rather than try to encourage a more active participation of cities and urban regions in the creation and execution of national development policies. The same is true for its predecessor, the Leipzig Charter (Zaucha 2011).

Rent-seeking is also a key challenge for multilevel horizontal governance as this behaviour creates negative externalities. Thus, as a counterbalance, the enhancement of synergies (positive externalities) is equally important. Markowski (2017) associates such positive externalities with the production of club goods and the diminishing of transaction costs. He provides an extensive discussion on the nature of externalities associated with urban centres. The Integrated Territorial Investment concept can be considered an attempt to address the dilemma of conscious territorial governance over externalities (with regard to cities and their surroundings). However, in practice, in many countries, it was used mainly as a vehicle for improving access to the EU funds rather than as a platform for enhancing a coordinated development within the urban regions.

Since, in many cases, a horizontal territorial governance has a voluntary character, a key prerequisite is the proper identification of interdependencies between cities and surrounding regions. This identification aids in determining who should be involved and for which type of policies. Here, the concept of FUA plays an important role. The majority of FUAs are constituted by functional relations, whereas the more formal governance is not adjusted to their territorial size and scope. Voluntary governance has some advantages and disadvantages. Its largest problem is that it is effective only with issues offering positive payoffs to all parties.

In general, multilevel governance and policy territorialisation play important roles in the ability of cities and urban regions to provide well-being and prosperity. The reason for their positive results is an efficient implementation of place-based policies that address local development assets and the creation of adaptive capacity at a local level (which is a key precondition for the proper functioning of a place-based policy paradigm). This kind of governance improves long-term development as well as economic and social resilience.

A multilevel governance (cooperation of cities) also creates a network-type of economies of agglomeration which are important for the development of medium-sized cities (Barca 2009, 18). It also enhances positive externalities and mitigates negative ones. All these factors increase the territorial competiveness, facilitate the functioning of the business sector and improve the quality of the citizens' lives. As pointed out by Markowski (2017: 71–72), 'In conditions of contemporary economy (i.e. a rapid decrease in internal competitive advantages of companies—achieved as a result of technical and organisational progress/…/) maintaining lasting competitive advantage of companies and territories is to an increasing extent determined by the quality of surroundings and regional (local) externalities cumulated in them'.

Multilevel governance requires proper territorial knowledge and territorial dialogue. The biggest risk lies in the lack of such knowledge, the inability to share the existing knowledge, limited know-how on the assessment of the territorial impacts of various policies, shallow and artificial territorial dialogue (Zaucha et al. 2017) as well as an aggressive influence of vested interests and politics on the dialogue (Markowski 2018). The possible solution can be an 'identification and incorporation of negative externalities into the social (electoral) system of values' but this is considered a long process requiring great mental shifts and conscious educational efforts (Markowski 2013, 62).

3.5 The Case of Polish Cities and Policy Territorialisation

Despite obvious centralisation attempts, Poland remains one of the pioneers in policy territorialisation. The territorialisation elements of development policy are included in the National Spatial Development Concept 2030. The existence of various forms of functional areas, including metropolitan areas and other urban functional areas, is foreseen there. In addition, urban areas constitute a significant part of functional areas identified by their structural problems (e.g. loss of industrial functions due to systemic transformation). In principle, a different policy and different intervention tools were to be addressed for particular types of territories.

In practice, functional areas have proven difficult to be delimited, so equipping them with territorially conscious policies is even more difficult. The aforementioned political changes towards centralization are not conducive to the empowerment of local units (including urban functional areas) as implementers of development policy. In spite of all these problems, the local dimension of social and economic problems is still noticed. The works carried out for the purposes of the new Responsible Development Strategy included the proposition (by Śleszyński et al. 2017a, b) of separately identifying areas of strategic state intervention (requiring support) and separately performing the function of growth poles (details of the separation—Bańskiet al. 2018; Komornicki et al. 2017). In the Strategy, 46 sub-areas threatened by marginalisation were identified in detail. The basis for the classification of territorial units in this category was their social and economic situation. In particular, the functional boundaries and location with respect to local and subregional cities—development centres—determined the boundaries of individual sub-areas. The adopted proposals assign a special role to a number of smaller cities that gather jobs and services of public benefit. Their share in the gross domestic product is usually small yet, at the same time, they are crucial for maintaining the standard of living in peripheral areas. According to the Strategy's principles, the horizontal cooperation of spatial functional units is a condition for obtaining external financial assistance (development aid).

There are also several practical examples related to policy territorialisation in the intraregional development policy in Poland both at a horizontal and a vertical level. The research conducted by the Institute for Development revealed more than twenty-five innovations related to multilevel governance and policy territorialisation (Zaucha et al. 2017, 219). For instance, in the Pomorskie Region, regional policy is partially implemented through integrated territorial agreements which are applied towards FUAs in this region. Those agreements are signed by the regional administration and a representation of various entities from a given FUA (local administration, educational sector, NGOs business sector) in order to secure finances and the implementation of an integrated development package encompassing various actions, including public investments. Those actions are then executed by various bodies participating in the agreement following the agreed upon commitments (Mikolajczyk 2015).

In the Zachodniopomorskie region, similar contracts are conducted with consortia of municipalities. Local representatives must form a partnership (with a minimum of three local governments plus the broad participation of non-public and public partners). The partnership should aim at solving a specific problem or stimulating development. The partnership should produce a diagnosis of the area covered by the agreement and prepare a strategy resulting in necessary investments. The advantage is that those investments will receive a privileged treatment (they can be financed outside call for proposals) under EU funds managed at a regional level. An additional incentive comes from the fact that the potential partnership consortia are aware that while only a limited number of contracts will be signed, they have unlimited freedom to form partnerships (Zaucha et al. 2017, 219).

In the Warminsko-Mazurskie region, the regional government has developed a way of encouraging the participation of local governments and other bodies from the region in international initiatives beneficial for the regional development. The main vehicle is information sharing. An example can be The Cittaslow initiative1 aiming at improving the quality of life in towns by slowing down its overall development pace, especially in a city's use of spaces and the flow of traffic through them. This deceleration will make cities more attractive for citizens and in-migrants. The regional government urged the urban administration to learn about the concept and analyse its potential effects for a given city or town. The financial support was offered to the towns and cities that voluntarily joined the movement (considering it beneficial for their development) only when international cooperation started producing visible effects. This condition was necessary in order to avoid the absorption bias as a reason for cities to join the initiative (Zaucha et al. 2017, 219).

In the Małopolskie region, the municipalities and counties were asked to voluntarily form planning and programming subregions. For such subregions, advisory and consultative bodies were established (subregional fora) to programme development policies and monitor territorial development, including the allocation of EU funds managed at a regional level. Subregional fora are composed of representatives of local governments (50% of participants) and social partners such as scientific communities, representatives of economy and civil society (non-governmental organisations), as well as public services such as Police, Fire Brigades and national parks headquarters.

Another useful example is the concept of the Regional Territorial Investment (RTI) employed in Poland. RTI is based on the ITI mechanism. Its uniqueness lies in its voluntary and bottom-up nature. Several Polish regions have decided to use such an instrument which allows for a supportive cooperation between local government units, not only in the largest metropolises but also in the FUA of subregional cities and towns. The RTI is a new territorial approach instrument in the regional development. It aims to inspire local governments to jointly implement a series of investments, creating a functionally coherent project.

The planned operations are implemented by either local governments or partnerships formed to tackle common challenges with the main requirement being the enactment of a collaborative vision of development. RTI projects are intended to be integrated into bundles so as jointly solve a development problem of a subregion. The 'Integrated Multifunction Passenger Exchange Node in Siedlce' can be considered a successful example of an RTI in the Mazowieckie Voivodeship. The project includes the expansion of the telecommunications infrastructure in the vicinity of the existing railway station in Siedlce, the construction of a bus interchange centre, the modernization of the telecommunications system in the adjacent districts and the establishment of the transfer centre. The highly positive evaluation of the project was based, among others factors, on the good level of cooperation between the partner cities and on the strong relationship between the project design and the strategic planning documents.

3.6 Territorial Utility with Focus on the Polycentricity Concept

It seems that the function of territorial utility was initially proposed by Mogiła (2015). However, similar ideas have also been recently discussed by other scholars (e.g. Nijkamp 2018) regarding a form of production function for city love. According to Mogiła (2017, 81), people derive utility from various sources including both consumption (access to jobs and other forms of income) and public goods provided in their area of residence and its spatial assets (i.e. spatial order, or the beauty of the landscape). Some of those assets are attached to the place of residence (e.g. spatial order) whereas some others can be reached by means of travelling (e.g. high-quality natural areas). Therefore, proximity of other 'attractive places' might also influence the territorial utility of neighbouring areas. Evident spillovers can be found in this case. Territorial utility, i.e. utility related to being a resident of a given place, can be revealed only indirectly (e.g. by examining individual location decisions of various social groups in various stages of their life cycle). However, territorial utility as such is a combination of market outcomes, which can be measured through consumption volume, and non-market decisions taken by public choice as it relates to the provision of public goods including maintenance of territorial assets such as spatial order, social capital or natural beauty. Consequently, a desirable level of territorial utility can be determined in the course of achieving social consensus, reflected then (or operationalised) in the policy of the public authorities. According to Mogiła's (2017, 81), public authorities should conventionally strive towards the maximisation of regional product (regional GDP) but at a given level of socially agreed territorial utility.

Mogiła's concept integrates three dimensions of the territorial cohesion outlined in Fig. 3.1. It takes into consideration territorial capital as a part of the regional production function, through which it supports the need for policy territorialisation, but it simultaneously refers to the importance of policy decisions in neighbouring areas (multilevel governance), and it finally integrates economic and territorial objectives within one utility function.

Although Mogiła does not directly associate this concept with cities, in his illustrative examples, cities do play a pivotal role. Using rigorous econometric models, Mogiła compares economic outcomes (in terms of GDP growth) of various types of public decisions related to the polycentricity level of the Dolnośląskie region in Poland.

Polycentricity, as a normative concept, is highlighted in key spatial-related documents produced within the European Union (ESDP, Territorial Agenda 2007 and 2011). The idea was initially created to counterbalance the dominance of the economic core of the European Union (the so-called Pentagon) and the largest cities located there (Kunzmann and Wegener 1991). This is also how it was understood in the first analysis of ESPON (ESPON 1.1.1 Final Report). Over time, the concept was also applied to smaller spatial units, including the ESPON 1.4.3 project, where its size was calculated for NUTS2 units. Polycentricity at lower geographical levels may sometimes raise doubts due to the lack of the possibility of creating new centres, as well as the threat of excessive fragmentation of development policy.

As underlined by Medeiros (2016, drawing on ESPON 2004), polycentricity has both morphological aspects, such as number of cities, connectivity, distribution and hierarchy and functional ones, such as flows, networks, cooperation, and functional complementarity. According to the definitions adopted by the ESPON 1.1.1 project (ESPON 1.1.1 Final Report 2003), there are three criteria for size (lack of dominance of one or several centres in terms of population or GDP), location (even distribution of the largest cities in the geographical space) and connectivity (existence of functional connections between most cities rather than only between one city and others). The fulfilment of these criteria must, however, be assessed by taking into account the size of the units under examination.

In the European scale (original ESPON study), countries such as Hungary are obviously less polycentric than Germany or Poland. The situation looks a bit different at the regional level. In the latter case, subnational indicators of polycentricity cannot be compared to those calculated for individual countries. Most often, the smaller the size of the region, the more monocentric it is. In general, polycentricity implies the strengthening of various types of cities/towns and their urban regions, the functioning of urban networks and the functional relations within functional areas (Zaucha and Szlachta 2017, 25). Mogiła's model covers both aspects.

The outcome of Mogiła's modelling that is relevant to policy is the estimation of the losses in GDP level (in comparison to non-polycentric patterns) that are necessary in maintaining various levels of polycentricity as an important territorial objective and the source of territorial utility. For instance, according to the Mogiła's Scenario No. 2, strengthening the attractiveness and agglomeration effects of the city of Wroclaw (regional capital and the largest city in the region) at the expense of other subregions around medium-sized cities would result in the increase of regional GDP by 8.1% in relation to the actual GDP of the year during which the calculation was performed (i.e. the GDP achieved under the actual polycentric pattern of the region). A substantial increase of resources (by 160%) in medium-size cities of the region and only by 5% in Wroclaw would increase this value to 9.9%, whereas modest concentration in the secondary cities and in Wroclaw (by 10 and 5%, respectively) would lower it to 4.2%. These figures illustrate the economic outcomes of various public choice decisions related to the regional polycentricity level. The actual choice belongs to the regional society represented by elected members of the regional assembly and their preferences for non-economic outcomes.

This dimension of territorial cohesion shows the new role of cities and urban regions. Its essence is in the creation of territorial utility for their citizens. This process encompasses the internal decisions of the cities with regard to their spatial objectives and the provision of public goods (such as spatial order, public transport, beauty of urban landscapes, local identity, social capital) and their role in supralocal networks that make decisions on necessary trade-offs between economic and spatial objectives. At first glance, this dimension may seem to be in conflict with the function of cities and urban areas as engines of growth. Yet, this is a superficial impression. Rather, one should see both as complementing one another, since economic growth and development should no longer be considered as the ultimate, overarching goal of all human actions regardless of its social, territorial and environmental outcome. The sources of human satisfaction and happiness are much broader.

3.7 The Case of Poland's Polycentricity

Poland belongs to the category of polycentric countries. The urban system is dominated by Warsaw but several other large cities, such as the Tri-city metropolis (Gdańsk-Gdynia-Sopot), Wroclaw, Poznań and Cracow, have demonstrated an outstanding performance in recent years. The spatial development of Poland is based on the desire to maintain this performance. Its policy is based on the notion of the Network metropolis (Korcelli et al. 2010), a key idea in the organisation of the national spatial policy. This notion has been officially introduced in the Polish National Spatial Development Concept, i.e. the national spatial development strategy adopted by the Polish Government in 2011 (Ministry of Regional Development (2011)), and has been implemented in a persistent way ever since.

To this day, almost all of the large Polish cities are connected via motorways and modernised railways shortening travelling time to the closest neighbours to below 3 h in one direction. However, at the same time, in Poland one can witness the diminishing role of the small towns that are any longer able to provide the necessary functions to their rural hinterlands. Churski and Hauke (2012) have diagnosed the significant relationship between the character of the urban settlement network and the polycentric development in Poland. 'The bigger the differentiation scale of national, regional, intraregional settlement patterns, and hegemony of the local centre, the smaller tendency towards polycentric development' (Churski and Hauke 2012, 93). However, they have also noticed that development and spatial policies mitigated polarisation and facilitated development of medium-sized cities in Poland.

Policies have a long-term nature. The condition for the final strengthening of the polycentric city system is the consistent continuation of policies that are territorially sensitive (not territorially indifferent). Despite the existence of a proper policy framework, in Poland this has resulted very challenging in practice. A perfect example is the investment policy in both road and rail transportation, along with the investment acceleration that took place after Poland's accession to the European Union. As a result of this investment, the territorial differentiation of the spatial accessibility indicators increased. In the field of transportation infrastructure, Poland has become more polarised. This is a natural phenomenon. The striking lack of spatial differences in accessibility prior to Poland's accession to the EU was the result of poor infrastructure in the entire country.

In general, the modernization of some routes leads to polarisation. However, this is not an irreversible process. Along with the development of the transportation network (motorways and expressways or modernised railway lines), a point is reached beyond which the disparities are once again reduced. Research conducted at the Institute of Geography and Spatial Organization of the Polish Academy of Sciences

(Komornicki et al. 2018) proves that, at the moment, Poland has already reached this break-even point in road transportation, while in the rail transportation such a point has not yet been attained.

If the investment process in Poland were to stop (e.g. due to lack of access to European Union funds in the next financial perspective after 2020), the country would have still benefitted from a transportation infrastructure that generally provides better access than in the previous 20 years. However, at the same time, this would further encourage spatial polarisation processes, in terms of accessibility, significantly more than before. This means the maintaining of a polycentric system at the level of the most important cities (the aforementioned network metropolis), while weakening it at the level of medium-sized cities of subregional importance.

The abovementioned considerations confirm that among the three polycentricity criteria, the most helpful actions are those supporting connectivity, broadly understood as both support for functional connections and infrastructure. At the same time, the size and location criteria change slowly. In fast-developing countries (like Poland), concentration processes are more likely to occur, which is a consequence of systemic transformation and the spatial polarisation of well-being, including the level of living and related conditions. Often, medium-sized cities (Śleszyński et al. 2017a, b) become the victims of this process, and thus polycentricity at the regional level. Cohesion policy and other activities with a territorial dimension (including support for the criterion of connectivity) are intended to counteract these negative tendencies.

In Poland, one can see the marginalisation of medium-sized cities, as well as some other regions, in the system of social (migration) as well as economic connections. Research (Komornicki et al. 2013) shows that the majority of migrants from rural areas and small towns move to either the capital or the largest provincial centres. In addition, many others migrate directly abroad. Only a few are fed into subregional centres (Fig. 3.3). The scale of migration between individual regional centres is also small, with the exception of those relating to Warsaw). This confirms the thesis that, in Poland, one can observe the disturbance of the polycentric settlement system, which becomes evident through its significant role in breaking the hierarchical system of internal migration.

Research on the populations of cities of different sizes (Śleszyński et al. 2017a, b) confirms these assumptions. They prove that both long-term socio-economic processes and new investment activities strengthen the polycentric system at the national level. In practice, the concentration of population (Fig. 3.4) and business activity takes place in several major cities (the aforementioned 'big five' and regional centres in eastern Poland), a phenomenon that weakens the position of most other cities, especially medium-sized ones.

Taking into account the research outcomes on interrelations between Polish metropolises, one can conclude that in Poland there is a supremacy of economic ties in comparison to an organisation of socio-economic systems and the dependency of other connections, especially demographic ones, on economic development. It is also assumed that one should be sceptical about limiting spatial polarisation by

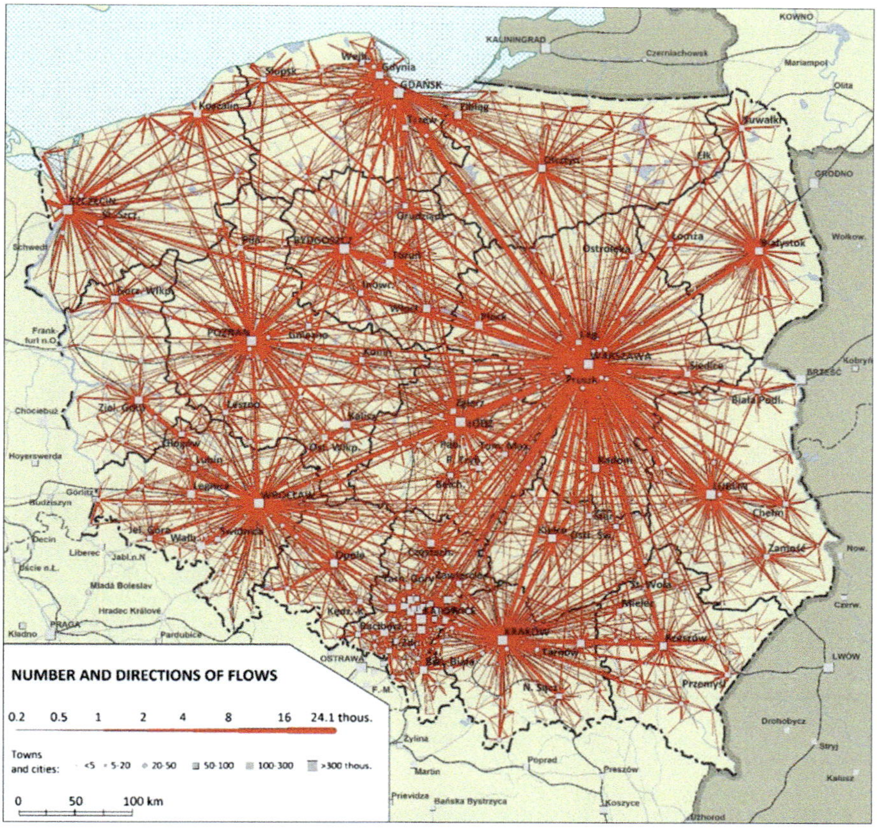

Fig. 3.3 Registered migration flows by current statistical reports 1989–2017. *Sources* Śleszyński P. based on the IGSO PAS material; Komornicki et al. (2013)

increasing the polycentricity level in all spatial scales. It has been suggested that processes in this field generally follow three potential scenarios (Komornicki et al. 2013 with modifications):

- Scenario I. Domination of the nation's capital. Transportation investments slow down. There is no development of multidirectional internal connections. Warsaw's domination grows in the metropolitan system. External relations are implemented mainly through the capital. Polycentricity is weakened in terms of the connectivity criterion, and most probably in other criteria, as well. Polycentricity at the regional level rapidly decreases, weakening the position of medium-sized cities.
- Scenario II. Network metropolis. Large infrastructure investments continue progressing. Multidirectional connections are strengthened. Gradual reduction of Warsaw hypertrophy towards other national metropolises is reduced. Polycentricity in terms of the connectivity criterion improves. In terms of size and location

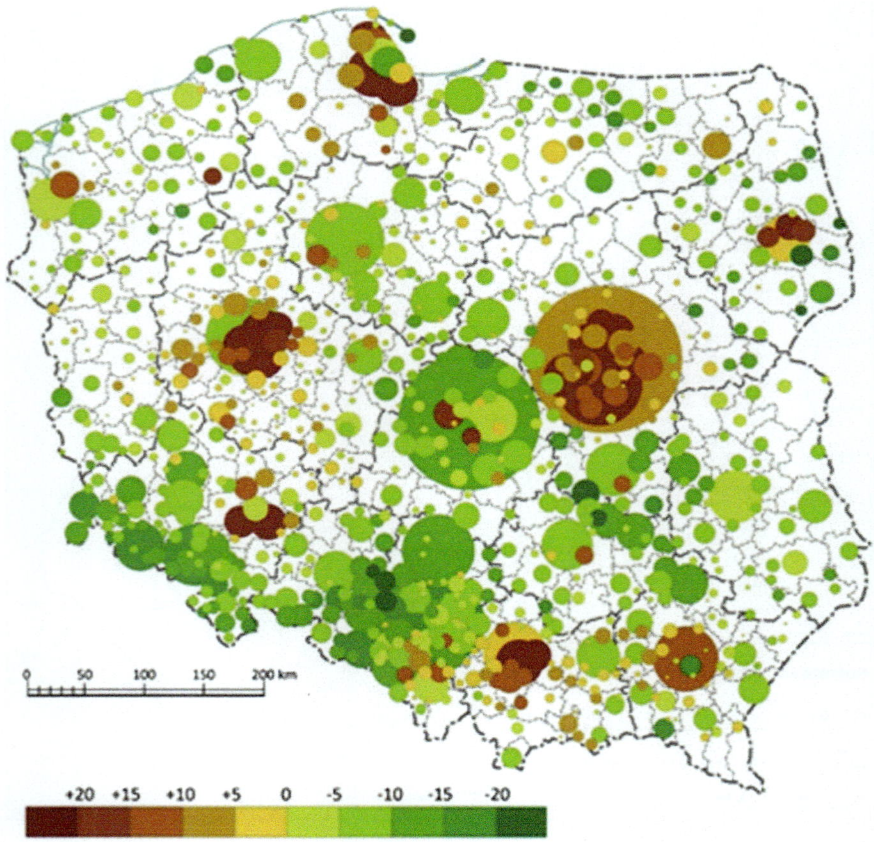

Fig. 3.4 Relative changes of the population size in % between 1998 and 2014 *Source* Śleszyński et al. (2017a, b), drawing on data of Statistics Poland Main Office—elaborated by M. Stępniak

criteria, negative trends are reversed at the national level, or at least slowed down at the regional level.

- Scenario III. Domination of foreign centres. Transportation investments slow down, while simultaneously the power of some centres from the 'adjacent metropolis' (foreign territories surrounding Poland) increases significantly. There is no development of multidirectional internal connections. The dominant position of Warsaw ceases to strengthen, but it is in favour of foreign metropolises. The level of polycentrism at a national scale does not change, but clearly decreases at the subregional level.

3.8 Conclusions

The analyses outlined above show that public authorities in various tiers, whose aim is to maintain the role of cities and urban areas as genuine engines of growth and welfare, should pay attention to all three dimensions of territorial cohesion. These three dimensions are tightly intertwined (Fig. 3.5). Their proper implementation, for the sake of an increase in the well-being and quality of life of its citizens, will require mutual understanding, coordination and co-operation from all stakeholders.

3.8.1 Territorial Capital: Economic Growth and Cohesion

The territorial capital of cities and urban areas is an important input in the local production function, a factor that conditions local resilience to external economic shocks and the main element of the lock-in effect (i.e. discouraging outflow of capital and labour force). Such capital should be built in a cohesive way for an entire functional urban region. Part of it is composed of non-economic assets, such as spatial order,

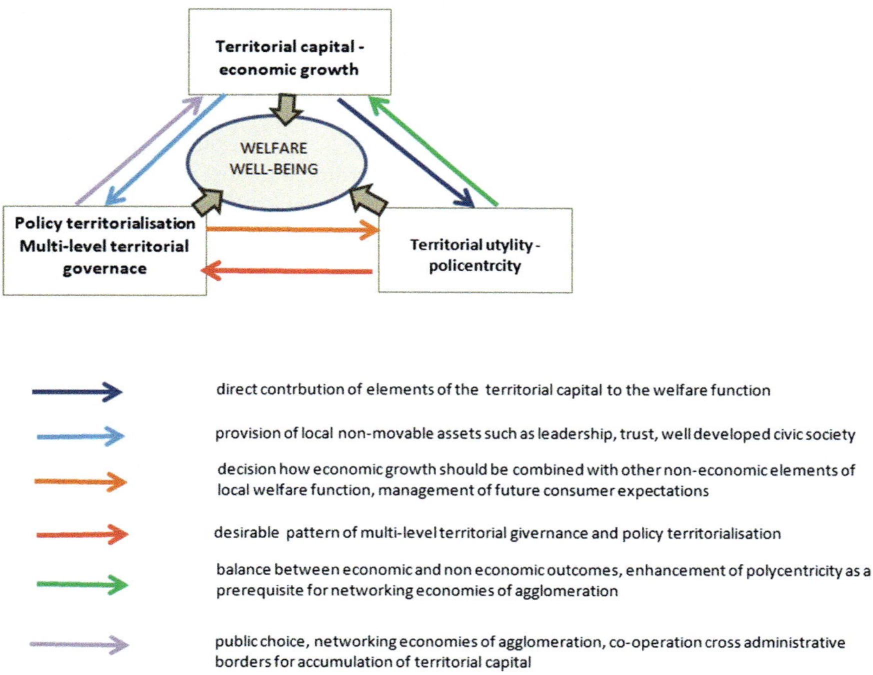

Fig. 3.5 Territorial cohesion in support of cities and urban regions delivering welfare and well-being. *Source* Own elaboration

landscape beauty and social capital. Such assets enhance total factor productivity and therefore contribute to economic growth. Territorial capital is directly linked to territorial governance (i.e. public choice) and territorial utility (i.e. balance between economic and non-economic outcomes, encouragement of polycentricity as a prerequisite for networking economies of agglomeration). Economies of agglomeration can be created by clustering as well as networking. Both benefit from multilevel territorial governance. Moreover, conscious accumulation of territorial capital requires the cooperation of various authorities as well as economic and social agents across formal territorial boundaries. This provides another link between territorial capital and territorial governance.

3.8.2 Territorialization of Policies: Multilevel Territorial Governance

The adjustment of size, scope and content of policies to the needs of various cities and urban areas is a *sine-qua-non* for the efficient use of local economic and institutional potential. Therefore, welfare is a co-product of place-based policies implemented in a dialogue between various territorial stakeholders. Each policy requires different territorial adjustments since each one should pay special attention to functional rather than formal regions and to functional interactions among such regions. Multilevel territorial governance is linked to territorial capital. It requires specific, local, non-movable assets such as leadership, trust and a well-developed civic society. It is also dependent on inputs from territorial utility. People's taste or opinion about the level of polycentricity influences public decisions regarding the most desirable territorial governance patterns. If those patterns do not match the stakeholder's preferences, they are less efficient, as a rule.

3.8.3 Territorial Utility: Polycentricity

Well-being and welfare should be considered as relative rather than absolute terms. Their scope and content depend on the values of the local societies, achieved level of development and agreed upon societal goals. They change with time and the dependent variables of citizens' utility function change accordingly. A city that functions, for a longer period of time, in line with the same mechanism that in the past has delivered welfare and well-being, may face malperformance in the next several years (e.g. Detroit). This calls for a constant process of rediscovery and/or reassessment of the welfare functions of inhabitants in a given city or urban region. This function guides the location decisions of the general public and the future of urban areas. Territorial utility is interlinked with territorial capital because elements of this capital represent a significant contribution to the welfare function. The beauty of the landscape or level of polycentricity directly affects the utility level of citizens. It is also connected to multilevel governance since, under the governance mechanism, a public choice decision must be made regarding how and to what extent economic growth should be combined with other non-economic elements of the local welfare function. Moreover, territorial utility heavily depends on the decisions of the adjacent territorial units. Thus, it requires inputs from external policies and their territorialisation. Territorial utility is influenced not only by the existing inputs but also by the expected ones. Therefore, policies, if properly adjusted to those needs, can influence the utility of consumers by encouraging the appropriate expectations.

3.8.4 Spatial and Economic Processes are Intertwined

Territorial cohesion cannot be implemented in a vacuum. One should not pursue economic policies without being aware of their territorial impact and vice versa. In addition, sometimes one should accept negative spatial outcomes while pursuing economic growth. However, these trade-offs should at least be clearly spelled out and consciously addressed. For instance, rapid economic development can take place at the expense of polycentricity, at least in the initial stage of growth (take-off stage). As has been shown in the Polish case, the concentration of GDP production in the largest urban centres is usually faster than the concentration of the population. Additionally, not all cities (including those existing up to now in the polycentric system) participate evenly in emerging network systems. Monocentricity can be strengthened in terms of the connectivity criterion, even if the other polycentricity criteria are equally good.

The response to economic concentration, shifted in time, is the migration from peripheral areas, usually directed almost exclusively at the strongest centres. It may lower the general level of polycentricity (especially at the regional level). In these situations, the task of the territorially sensitive development policy is mainly to mitigate the described trends. The efforts to achieve the maximum in terms of polycentric development should be undertaken. In particular, policy should pay attention to strengthen the criterion of connectivity (continuing the territorially sensitive investment policy while maintaining its appropriate coordination with other policies). In the course of time and with an increase in the level of well-being, perhaps the economic concentration trends will be reversed by people's spatial decisions, companies induced to decrease communication and transportation costs, as well as an increased importance of territorial capital as predicted by the New Economic Geography.

Acknowledgements The chapter presents the outcomes of the research by the Institute of Development in Sopot financed by the Polish National Science Centre under a project: 'Concept of the territorial cohesion in cohesion policy. Implications for Economic Growth' (no. 2012/05/B/HS4/04212).

References

Arouri M, Youssef AB, Nguyen-Viet C, Soucat A (2014) Effects of urbanization on economic growth and human capital formation in Africa. Program on The Global Demography of Aging at Harvard University Working Paper No. 119. Retrieved from http://www.hsph.harvard.edu/pgda/working.htm. Access 20 May 2018

Baldwin R, Okubo T (2009) Tax reform, delocation, and heterogeneous firms. Scand J Econ 111(4):741–764

Bański J (2008) Przemiany funkcjonalno-przestrzenne terenów wiejskich. In: Ekspertyzy do Koncepcji Przestrzennego zagospodarowania Kraju 2008–2033. Tom I, Warszawa: Ministry of regional development, pp 393–434 [in Polish]

Bański J, Degorski M, Komornicki T, Śleszyński P (2018) The delimitation of areas of strategic intervention in Poland: a methodological trial and its results. Moravian Geogr Rep 26(2):84–94

Barca F (2009) An agenda for a reformed Cohesion Policy. A place-based approach to meeting European Union challenges and expectations Independent Report prepared at the request of Danuta Hübner, Commissioner for Regional Policy, available at: http://ec.europa.eu/regional_policy/archive/policy/future/barca_en.htm

Bottazzi L, Peri G (2003) Innovation and spillovers in regions: evidence from European patent data. Eur Econ Rev 47(4):687–710

Brodzicki T (2018) The role of openness in economic growth of regions. The perspectives of economic growth and new economic geography theory and empirics. Working paper of Institute for Development 001/2018 (040)

Brülhart M, Sbergami F (2009) Agglomeration and growth: cross country evidence. J Urban Econ 65(1):48–63

Camagni R (2008) Regional competitiveness: towards a concept of territorial capital. In: Capello R, Camagni R, Chizzolini B, Fratesi U (eds) Modelling regional scenarios for the enlarged Europe. European Competiveness and Global Strategies, Springer, Berlin, pp 33–49

Churski P, Hauke J (2012) Polycentric development and growth areas (w:) T. Markowski, M. Turała (red.) Territorial Cohesion Policy in Poland – Issues In Impact Assessment. Studia Regionalia KPZK PAN, vol. 33, Warsaw: KPZK PAN, pp 79–96

Cieślik A, Rokicki B (2013) Rola sieci transportowej w rozwoju polskich regionów: zastosowanie modelu potencjału ekonomicznego. Acta Universitatis Niciolai Copernici, Ekonomia XLIV(1):113–127

Ciołek D (2017) Oszacowanie produktu krajowego brutto w polskich powiatach [Estimation of gross domestic product in Polish Local Administrative Districts]. Gospodarka Narodowa 3(289):55–87

Ciołek D, Brodzicki T (2015) Territorial capital and polish regional development: a neoclassical approach. Institute for Development Working Papers no. 011/2015

Ciołek D, Brodzicki T (2016) Determinanty produktywności polskich powiatów [Productivity determinants of Polish Local Administrative Districts]. Bank i Kredyt 47(5):463–494

Ciołek D, Brodzicki T (2017a) Spatial dependence structure of total factor productivity in polish local administrative districts. Acta Universitatis Lodziensis. Folia Oeconomica 3(329):73–92

Ciołek D, Brodzicki T (2017b) Territorial capital and polish regional development: a neoclassical approach. In: Bradley J, Zaucha J (eds) Territorial cohesion: a missing link between economic growth and welfare. Lessons from the Baltic Tiger". Gdańsk: University of Gdańsk and Institute for Development, pp 147–168

Colenbrander S (2016) Cities as engines of economic growth: the case for providing basic infrastructure and services in Urban areas. IIED Working Paper

Davoudi S, Zaucha J, Brooks E (2016) Evolutionary resilience and complex lagoon systems. Integr Environ Assess Manag 12(4):711–718

Dixit A (2009) Governance, institutions and economic activity. Am Econ Rev 99(1):5–24

Doucet P, Böhme K, Zaucha J, (2014) EU territory and policy-making: from words to deeds to promote policy integration. Debate article. European Journal of Spatial Development

Du D (2017) The causal relationship between land urbanization quality and economic growth. Qual Quant 51(6):2707–2723
Duranton G (2014) The urbanization and development puzzle. In Shahid Y (ed) The Buzz in cities: new economic thinking. Washington D.C., The Growth Dialogue, pp 1–19
EC (2017) My Region, My Europe, Our Future – Seventh report on economic, social and territorial cohesion. Luxembourg: office for official publications of the european communities
Ellison G, Glaeser EL, Ker WR (2010) What causes industry agglomeration? Evidence from coagglomeration patterns. Am Econ Rev 100:1195–1213
ESPON (2004) Potentials for polycentric development in Europe. Luxembourg: ESPON Project 1.1.1. Luxembourg: ESPON
ESPON (2013) ESPON TANGO—Territorial Approaches for New Governance. Scientific Report Version 20/12/2013. Luxembourg: ESPON and Nordregio
Faludi A (2015) Place is no man's land. Geographia Pol 88(1):5–20
Faludi A (2016) EU territorial cohesion, a contradiction in terms. Plann Theor Pract 17(2):302–313
Forslid R, Okubo T (2010) Spatial relocation with heterogeneous firms and heterogeneous sectors. CEPR Discussion Paper No. DP8117
Glaeser EL (2013) A World of Cities: the Causes and Consequences of Urbanization in Poorer Countries, NBER Working Paper 19745
Henderson V (2003) The urbanization process and economic growth. The so-what question. J Econ Growth 8:47–71
Komornicki T (2013) Assesment of Infrastructure construction, its role in regional development, report. GRINCOH Working Paper 604. Retrieved from http://www.grincoh.eu/media/serie_6_spaces__territories_and_regions/grincoh_wp6.04_komornicki.pdf. Access 10 Aug 2018
Komornicki T (2016) Migration outflow from rural areas in Poland after the year 2004, Landflucht? Gesellschaft in Bewegung. *Informationen zur Raumentwicklung*, Heft 2.2016, Bundesinstitut fur Bau-, Stadt- und Raumforschung, pp 151–160
Komornicki T, Ciołek D (2017) Territorial capital in Poland. In: Bradley J, Zaucha J (eds) Territorial cohesion: a missing link between economic growth and welfare. Lessons from the Baltic Tiger. Gdańsk: University of Gdańsk and Institute for Development, pp 93–146
Komornicki T, Czapiewski K (2016) Regional development in Świętokrzyskie and Podkarpackie Regions. Background analysis for the project, polish lagging regions. Warszawa, World Bank
Komornicki T, Korcelli P, Siłka P, Śleszyński P, Świątek D (2013) Powiązania funkcjonalne pomiędzy polskimi metropoliami. Wydawnictwo Akademickie SEDNO, Warszawa, p 218
Komornicki T, Rosik P, Śleszyński P, Stępniak M (2014) Raport przygotowany w ramach III etapu projektu: Oszacowanie oczekiwanych rezultatów interwencji za pomocą miar dostępności transportowej dostosowanych do potrzeb dokumentów strategicznych i operacyjnych dotyczących perspektywy finansowej 2014–2020. Ministerstwo Inwestycji i Rozwoju, Warszawa [in Polish]
Komornicki T, Bański J, Degórski M, Śleszyński P (2017) Obszary Wzrostu jako przedmiot strategicznej interwencji państwa. In: Okrasa W (ed) Jakość życia i spójność przestrzenna: Interakcje rozwoju i dobrostanu w kontekście lokalnym. War5szawa: Wydawnictwo naukowe UKSW, pp 157–176
Komornicki T., Rosik P., Stępniak M., Śleszyński P., Goliszek P., Pomianowski W., Kowalczyk K.(2018) Evaluation and monitoring of accessibility changes in Poland using the MAI indicator, IGSO PAS, MED., Warsaw, p 92
Korcelli P, Degórski M, Drzazga D, Komornicki T, Markowski T, Szlachta J, Węcławowicz G, Zaleski J, Zaucha J (2010) Ekspercki projekt koncepcji przestrzennego zagospodarowania kraju do roku 2033 [Expert proposal for the National Spatial Development Concept till the year 2033]. *Studia KPZK PAN*, 128
Kunzmann KR, Wegener M (1991) The Pattern of Urbanisation in Western Europe 1960–1990. Berichte aus dem Institut für Raumplanung 28. Dortmund: Institut für Raumplanung, Universität Dortmund

Markowski T (2013) Territorial dimension of integrated development policy—expectations and challenges concerning planning and institutional systems. Studia Regionalia KPZK PAN 35:51–64
Markowski T (2017) New externalities in development of creative and intelligent city—theoretical concept. Studia Regionalia KPZK PAN 51:69–81
Markowski T (2018) Polityka urbanistyczna państwa—koncepcja, zakres i struktura instytucjonalna w systemie zintegrowanego zarządzania rozwojem [State Urban Policy—concept, Institutional scope and structure in integrative management of the development]. Studia KPZK PAN 183:89–101
Marshall A (1879) Economics of industry. Macmillan, London
Marshall A (1890) Principles of economics. Macmillan, London
Medeiros E (2014) Territorial Impact Assessment (TIA). The process, methods and techniques. Lisbon, Centro de cEstudios Geográficod da Universidade de Lisboa—Instituto de Geografia e Ordenamento do Território
Medeiros E (2016). Territorial cohesion: an EU concept. Eur J Spat Dev, 60. Retrieved from https://relocal.eu/wp-content/uploads/sites/8/2017/01/Refereed_60-1.pdf. Access 17 May 2018
Mikolajczyk A (2015) *Pomorskie doświadczenia dialogu terytorialnego jako przykład dobrej praktyki w zakresie realizowania terytorialnie ukierunkowanej polityki rozwoju* [Experience of Pomerania Region on territorial dialogue as a good practice related to place-based evelopment policy]. In: Zaucha J, Brodzicki T, Ciołek D, Komornicki T, Mogiła Z, Szlachta J, Zaleski J (2015) Terytorialny wymiar wzrostu i rozwoju. Warszawa, Difin, pp 396–406
Mills E, McDonald J (2017) Sources of metropolitan growth. Routledge, London, New York
Ministry of Regional Development (2011) National spatial development concept. Ministry of Regional Development, Warsaw
Mogiła Z (2015) Conceptual model of the concept of the territorial cohesion. Working paper of Institute for Development 010/2015 (029)
Mogiła Z (2017) Cohesion as a territorial optimum. In: Bradley J, Zaucha J (eds) Territorial cohesion: a missing link between economic growth and welfare: lessons from the Baltic Tiger. Gdańsk: University of Gdańsk and Institute for Development, pp 73–92
Moreno R, Paci R, Usai S (2005) Spatial spillovers and innovation activity in European regions. Environment and Planning A, 37:1793–1812
Mukhopadhyay P (2018) Engines without drivers: cities in India's growth story. CPRINDIA Working Paper. Retrieved from http://www.cprindia.org/research/papers/engines-without-drivers-cities-india%E2%80%99s-growth-story. Access 21 July 2018
Nijkamp P (2010) Regional development in an unstable word. In. Churski P, Ratajczak W (eds) Regional development and regional policy in Poland: first experiences and new challenges of the European Union membership. Studia Regionalia KPZK PAN 27, part 1: 10–18
Nijkamp P (2018) The body and soul of space: a microcosmic perspective. Presentation at the conference spatial development of Poland in the face of civilization challenges of the Committee for Spatial Economy and Regional Planning, Polish Academy of Sciences, Jachranka 5–6 June 2018. Retrieved from http://kpzk.pan.pl/images/prezentacje_kpzk60/Peter-Nijkamp-Conference-Warsaw3.pdf. Access 20 July 2018
North D (1990) Institutions, institutional change, and economic performance. Cambridge University Press, Cambridge
OECD (1997) Regional competitiveness and skills. OECD, Paris
Ottaviano GM (2010) 'New'new economic geography: firm heterogeneity and agglomeration economies. J Econ Geogr 11(2):231–240
Rodríguez-Pose A (2013) Do institutions matter for regional development? Reg Stud 47(7):1034–1047
Rodríguez-Pose A, Ketterer T (2016) Institutional change and the development of lagging regions in Europe. Retrieved from https://az659834.vo.msecnd.net/eventsairwesteuprod/production-ersa-public/1a1b44496f22469194d50076e57fbbe4. Access 21 May 2018

Rosik P, Stępniak M, Komornicki T (2015) The decade of the big push to roads in Poland: impact on improvement in accessibility and territorial cohesion from a policy perspective. Transp Policy 37:134–146

Śleszyński P (2013) Delimitacja Miejskich Obszarów Funkcjonalnych stolic województw. Przegląd Geograficzny 85(2):173–197 [in Polish]

Śleszyński P, Bański J, Degórski M, Komornicki T (2017) Delimitacja Obszarów Strategicznej Interwencji Państwa: obszarów wzrostu i obszarów problemowych, 2017, Warszawa, Instytut Geografii i Przestrzennego Zagospodarowania PAN, 310 p (Prace Geograficzne, 260)

Śleszyński P, Bierzyński A, Cerić D, Ciechański A, Degórska B, Degórski M, Komornicki T, Korcelli-Olejniczak E, Piotrowski F, Stępniak M, Węcławowicz G, Wiśniewski R (2017b) Analiza sytuacji rozwojowej miast, w tym charakterystyka miast tracących funkcje społeczno-gospodarcze, Raport dla Ministerstwa Rozwoju. Ministerstwo Rozwoju, Warszawa

Smętkowski M, Jałowiecki B, Gorzelak G (2008) Obszary metropolitalne w Polsce: problemy rozwojowe i delimitacja, *Raporty i Analizy EUROREG, 1* [in Polish]

Spiekermann K, Wegener M, Květoň V, Marada M, Schürmann C, Biosca O, Ulied Segui A, Antikainen H, Kotavaara O, Rusanen J, Bielańska D, Fiorello D, Komornicki T, Rosik P, Stepniak M (2013) TRACC transport accessibility at regional/local scale and patterns in Europe. Final Report. ESPON Applied Research

Staniek Z (2007) Zróżnicowanie ekonomii instytucjonalnej [Differences in institutional economics]. Retrieved from http://www.e-sgh.pl/niezbednik/student_przedmiot.php?pid=1081. Access 29 May 2018

Szlachta J, Zaucha J, Komornicki T (2017) Polish development policy and its territorial dimension. In: Bradley J, Zaucha J (eds) Territorial cohesion: a missing link between economic growth and welfare. Lessons from the Baltic Tiger. Gdańsk, University of Gdańsk and Institute for Development, pp 49–71

World Bank (2009) World development report 2009: reshaping economic geography. World Bank, Washington D.C

World Bank (2015) Competitive cities for jobs and growth. World Bank Group, Washington D.C.

World Bank (2017) Cities in Europe and Central Asia: Poland. World Bank Group, Washington, D.C

Yusuf, S., Leipziger, D. (2014) Making city lights shine brighter. *CEPR Policy Insight*, 71

Zaucha J (2007) Rola przestrzeni w kształtowaniu relacji gospodarczych—ekonomiczne fundamenty planowania przestrzennego w Europie Bałtyckiej [Territory as a factor determining economic relations—economic foundations for spatial planning in the Baltic Sea Region]. Fundacja Rozwoju Uniwersytetu Gdańskiego, Gdańsk

Zaucha J (2011) Spatial functional linkages according to the economist. In: Komornicki T, Siłka P (eds) Functional linkages between polish metropolises. Studia Regionalia KPZK PAN 29: 130–147

Zaucha J (2014) Regional institutions and economic growth of regions. Working paper of Institute for Development 002/2013 (013)

Zaucha J, Szlachta J, (2017) Territorial cohesion: origin, content and operationalization. In: Bradley J, Zaucha J (eds) Territorial cohesion: a missing link between economic growth and welfare. Lessons from the Baltic Tiger. Gdańsk: University of Gdańsk and Institute for Development, pp 23–47

Zaucha J, Świątek D, Stańczuk-Olejnik K (2013) Place-based territorially sensitive and integrated approach. Ministerstwo Infrastruktury i Rozwoju, Warszawa

Zaucha J, Komornicki T, Böhme K, Świątek D, Żuber P (2014) Territorial keys for bringing closer the territorial agenda of the EU and Europe 2020. Eur Plan Stud 22(2):246–267

Zaucha J, Komornicki T, Szlachta J (2017) Territorial cohesion in Polish region al development Policy as a process of adaptation to the specifics of a territorial unit. In: Bradley J, Zaucha J (eds) Territorial cohesion: a missing link between economic growth and welfare. Lessons from the Baltic Tiger. Gdańsk: University of Gdańsk and Institute for Development, pp 205–222

Jacek Zaucha holds a Ph.D. in economics, member of the faculty of the Economics Department of the Gdańsk University (since 1994), also associated with the Maritime Institute in Gdańsk (since 2007), the former member of the Committee on the Spatial Development in the Baltic Sea Region-VASAB 2010, member of the Drafting Team elaborating the 'Concept of the Spatial Development of Poland', member of the drafting team for the update of 'The Territorial State and Perspectives of the European Union' document and the revision of the 'Territorial Agenda of the EU', one of the key authors of the report of Polish Presidency of EU in the field of territorial cohesion, author of 115 scientific publications on: supranational spatial planning and development, EU structural/cohesion policy, spatial integration and co-operation.

Tomasz Komornicki is a Professor born in 1963. A graduate of the Warsaw University (1988), Ph.D. in Polish Academy of Sciences (1998), Head of the Department of Spatial Organization in the Institute of Geography and Spatial Organization Polish Academy of Sciences (PAS), Professor at the Faculty of Earth Sciences and Spatial Management, Maria Curie-Sklodowska University in Lublin; member of the executive body of the Committee for Spatial Economy and Regional Planning, PAS, head and participant of many Polish and international research projects, including HORIZON, ESPON and INTERREG projects; member of the Consulting Board for preparation of the new Spatial Development Concept of Poland up to 2030; member of the international scientific team preparing the Territorial Agenda of the European Union 2020. He authored more than 400 scientific publications. His main areas of interest are geography of international interactions, transport geography and spatial planning.

Part II
Territorial Cohesion and Urban Balanced Systems?

Chapter 4
Polycentric Urban Systems and Territorial Cohesion

Jiří Malý

Abstract This chapter outlines the idea of polycentric development, one of the core principles of territorial cohesion discourse. As a normative and ideological vision of balanced spatial development, polycentric urban systems are seen as effective spatial arrangements overcoming regional disparities. Commonly shared policy imaginaries of polycentricity, however, have yet to fully reflect the complex nature of urban system dynamics. The chapter offers deeper insight into the three crucial facets of polycentricity: scale-dependency, normativity versus factual practices, and co-operative relationships, which together contribute to the ambiguous and multi-interpretative character of polycentric development. As a scale-dependent concept, polycentricity has various forms and implications for urban systems at different geographical scales. Consequently, universal applicability of this normative concept is hampered by colourful urban spaces and overwhelming territorial diversity. Even if fundamental spatial principles of polycentricity are fulfilled, existing territorial settings and urban relations may diverge from cohesive spatial patterns and lead to increased urban competition. Finally, co-operative relationships among urban centres within polycentric spatial structures are not fully developed because of excessive institutional fragmentation and a missing framework for tackling collectively shared problems. These issues are discussed within a broader theoretical debate on urban systems and territorial cohesion while referring to the essential studies which serve as evidence of the intricate all-encompassing analytical grasp of polycentricity.

Keywords Polycentricity · Urban systems · Territorial cohesion
Scale-dependency · Co-operation

J. Malý (✉)
Czech Academy of Sciences, Institute of Geonics, Drobného 28, 602 00 Brno, Czech Republic
e-mail: jiri.maly@ugn.cas.cz

E. Medeiros (ed.), *Territorial Cohesion*, The Urban Book Series,
https://doi.org/10.1007/978-3-030-03386-6_4

4.1 Introduction

One of the main objectives of the European Union (EU) territorial cohesion policy agenda is to promote polycentric and balanced spatial development across the European territory (EC 1999; Territorial Agenda 2007, 2011; ESPON 2014). Since the adoption of the European Spatial Development Perspective (ESDP), polycentric territorial settings as well as fair access to infrastructure, public amenities, and knowledge have become the cornerstones of the long-term EU cohesion strategy.[1] As an umbrella concept, territorial cohesion has an undoubtedly multi-interpretative character which results in a number of mutually competing narratives and storylines (Waterhout 2007; Evers et al. 2009). A dual interpretation of territorial cohesion, which consists of a narrative emphasizing harmonic and balanced development and a narrative stressing support for competitiveness and economic growth, has become the prevailing one (Ache et al. 2008; Vanolo 2010). The objective of territorial cohesion in covering aspects of social solidarity and the welfare state may seem like a contradiction to neoliberal principles that operate on the basis of the free market and the full utilization of competition mechanisms in today's globalized environment. However, within the framework of European policies, territorial cohesion and competitiveness are relatively strongly connected, despite the fact that the conceptualization of this relationship is not clear (Vanolo 2010). In this context, polycentric development is seen as a tool bridging the cohesion/competitiveness discourses as well as ensuring the balanced development and economic growth of the territories concerned.

This chapter gives an overview of the issues related to the concept of polycentricity and its importance in achieving territorial cohesion policy goals (see Medeiros 2016). Two main perspectives of the multi-dimensional concept of polycentricity—structural and institutional—provide a basic conceptual framework for the following discussion of the three fundamental facets of polycentricity, which deserve special attention from urban planners and relevant policymaking authorities. First, as a genuine spatial concept, polycentricity requires different approaches and interpretations dependent on the scale being considered. Reflecting the scale-related interdependencies, the analytical treatment of polycentricity ranges from the evaluation of international territorial disparities to analyses of cross-border relations, inter-municipal co-operation processes, and the identification of urban sub-centres. Second, normative assumptions about polycentricity are tested by examining factual spatial practices; however, empirical evidence of the expected benefits generated in polycentric spatial structures is mostly missing. Third, attention is paid to the establishment of co-operative relationships in polycentric urban systems, which are often neglected in polycentricity-oriented studies but are certainly amongst key elements in the effective performance of polycentric urban regions. The ambition of this chapter is not to cover all aspects of the multi-dimensional concept of polycentricity but to discuss

[1] Supporting balanced polycentric urban systems and a new relationship between cities and rural areas became the key principles of the ESDP at a meeting of ministers responsible for spatial planning in Leipzig in 1994 (ESPON SPON 2005).

current shortcomings of the polycentricity discourse in order to provide a springboard for future research and point out research streams and topics which deserve further attention.

4.2 Polycentricity as a Normative Concept: Structure and Governance

Rooted in the French tradition of 'l'aménagement du territoire', the interpretation of territorial cohesion, as a vehicle to promote a balanced Europe, is characterized by strong support for policy cohesion objectives and redistribution of financial resources so as to reduce regional disparities (Faludi 2004). This view stems from the perception that globalization and European liberal economies are seen as a threat to regional and spatial development, meaning that only the strongest regions and localities will be favoured while other peripheral locations further decline socially and economically (Robert 2007). Territorial cohesion is here perceived from the traditional perspective of economic and social cohesion, which aims at moderating interregional inequalities, and is associated with the development of mostly peripheral and more disadvantaged regions characterized by poor accessibility and low population densities. Balanced territorial development is the prerequisite for the subsequent competitiveness of Europe as a whole. On the other hand, the narrative of economic competitiveness understands globalization as a phenomenon that should be used to profit Europe and its economic performance in order to keep up with global competition. Specifically, under this economic competitiveness rationale, investments should give preference to the most developed urban cores. From this stance, the European territorial structure would be better optimized; the aim is not only to attain global competitiveness of Europe as a whole but also the competitiveness of individual regions. Ensuring competitiveness sustains the achievement of territorial cohesion goals, as demonstrated by the commonly agreed EU cohesion strategy which expressed that investment in regional competitiveness improves people's lives and is the basis for growth in less developed regions (ESPON 2014).

Political normative narratives underlying the territorial cohesion discourse have been transformed into specific spatial planning visions, images and concepts, including the rationale for promoting polycentric development. As a result of political debates, polycentricity is understood as a normative spatial planning concept with the potential to better manage the growing complexity of urban systems. Furthermore, polycentric development should be able to bridge the fundamental cohesion and competitiveness dichotomy which is emblematic of the territorial cohesion discourse. In line with the EU cohesion narrative, polycentricity is associated with equal access to services of general interest and jobs, i.e. the basic precondition for territorial cohesion established by the Treaty of Amsterdam (EC 1997). Polycentric development represents here the 'fairness' principle of the territorial cohesion discourse expressed by the socially motivated attitude towards mitigating regional disparities and ensuring social solidarity and spatial justice within capitalist economies subject to market forces (Davoudi 2005; Malý 2018). In terms of a competitive Europe, Waterhout (2008: 108) claims that 'the clearest example, though, of competitiveness being taken into account is the concept of global economic integration zones'. From the perspective of polycentric development, these zones should form a counterweight to the European economic core in order to ensure more balanced regional development. In this context, a competitive Europe is perceived as a precondition which builds on territorial potentials, smart growth, and integrated economic centres that take the form of polycentric spatial structures. Balanced spatial development would then be achieved as a result of diffusion processes (Böhme and Gløersen 2011).

The concept of polycentricity has been discussed both from the perspective of structural preconditions, stemming from the spatial configurations of human activities, and that of institutional frameworks. While the former perspective is focused especially on the morphological and functional structures of national/regional/urban spatial arrangements which have to fulfil some predefined criteria in order to be considered polycentric (Kloosterman and Musterd 2001; Parr 2004), the latter emphasize new forms of territorial governance, multilevel decision-making, public-policy interventions, and co-operation between various stakeholders (Albrechts 2001; Finka and Kluvánková 2015). The structure of urban spatial configurations is at the forefront of the strategic and urban planning apparatus, and it is approached analytically and evaluated by using specific spatial and statistic algorithms. Polycentricity, as an institutional process, refers to institutional integration and political co-operation within polycentric spatial arrangements, which belong to topics discussed within political theories and policy analyses.

The structure of polycentric urban systems is mostly related to the balanced distribution of human activities across urban space that is organized non-hierarchically (Kloosterman and Musterd 2001). Polycentric urban systems are characterized by the absence of one dominant urban node and the presence of rather similar-sized centres which form urban networks with urban centres that mutually interact and establish collaborating spatial units. In the context of urban system theories, we can observe a shift from traditional central place models (Christaller 1933) and hierarchical models (Berry 1964), which were unable to describe the new spatial-functional organization of urban systems that appeared together with the rise of globalization and the post-industrial economy, towards network and polycentric models (Camagni 1993; Parr 2004). These are attempts to analytically grasp the emerging spaces of flows and increasing complexity of human interactions. Polycentric development has become a widely discussed issue with regard to its expected ability to gain benefits: positive externalities and economies of scale exhibited in large agglomerations while avoiding undesirable agglomeration diseconomies and urban sprawl typical of monocentric spatial structures (Bailey and Turok 2001).

Two basic approaches to defining (and measuring) polycentricity have been established: morphological and functional. The morphological approach emphasizes size (e.g. population, number of jobs, etc.) and distribution of cities across space. Morphological polycentricity thus points to non-hierarchical urban systems, where larger numbers of urban centres are situated in close proximity, and where human activities are evenly spread across the territory. The functional approach stresses the importance of the relational dimension of urban systems. As such, functional polycentricity represents balanced and reciprocal functional linkages which prevail in a particular urban system. Instead of flows directed at one strong centre, functionally polycentric urban systems comprise flows creating mutual and criss-crossing interactions (de Goei et al. 2010). Urban centres complement each other in terms of higher ranked urban functions and services. Morphological and functional approaches should be implemented together as the size and distributional polycentric configurations do not necessarily guarantee the functional interconnectedness of a particular urban system (Albrechts 2001).

While a plethora of studies focusing on conceptual and analytical issues have been published, polycentricity, as a structural process, has reached a commonly shared and accepted theoretical framework and has been conceptually anchored (Burger and Meijers 2012; Malý 2016a). Conversely, the issue of public-policy interventions that promote polycentric urban systems has not yet been fully examined, at least in relation to the territorial cohesion discourse. The anticipated benefits of polycentric spatial structures are fulfilled only when joint coordination of institutions, public policies, and other actors involved have been established. Polycentric governance refers to multiple governing authorities at differing spatial scales which are joined into 'contractual and co-operative undertakings or have recourse to central mechanisms to resolve conflicts' (Ostrom et al. 1961: 831). A polycentric governance system is broadly understood as a platform where each institution has substantial independence to enforce rules within a circumscribed scope of authority for a specified spatial area (Ostrom 1999). A polycentric institutional framework may include general-purpose governments as well as highly specialized units (McGinnis 1999).

In understanding the structural configurations of polycentricity as a precondition for the application of polycentric governance, we cannot avoid one fundamental interdependency: the inclusion of polycentric governance within the same spatial area in which the polycentric urban system structurally operates. Nevertheless, the institutional framework of public policies is mostly tied to territorial units based on administrative settings and does not usually correspond to the spatial-functional arrangements of urban systems. Increased mobility, migration, the dynamics of economic processes, and the growing unpredictability of human behaviour affect the territorial grounding of human activities and the overall sense of space and place.

Administrative borders are no longer an appropriate form for the delimitation of territorial units used for spatial planning purposes, let alone governance processes for implementing polycentric spatial structures. Indeed, the borders of urban systems are becoming increasingly blurred and fuzzy (Finka and Kluvánková 2015). Hence, the territorial cohesion discourse requires rethinking traditional territorial settings, including administrative and institutional hierarchies, so that the vision for a polycentric urban system is not a mere abstract illusion. While the structural preconditions of urban systems show some degree of stability and the main socio-economic linkages often occur in similar functionally coherent spatial arenas (e.g. financial bonds between global cities, flows of goods and capital between urban agglomerations, daily routine work and leisure commuting), institutional frameworks should be able to respond to unusual situations stemming from the dynamic of contemporary social and environmental processes in order to avoid rigid public administration processes and to incorporate the principles of multilevel governance. Further interpretation of polycentric urban systems (both in terms of structure and governance), however, calls for the discussion of various spatial scales and hierarchical levels where the concept of polycentricity could be applied.

4.3 Scale-Dependency of Polycentricity

Conceptual and analytical treatment of polycentricity faces the heterogeneity of spatiality, especially with regard to the applied territorial scale. The territorial Agenda of the EU (2020) suggests that polycentric development must be pursued at the macro-regional, cross-border, as well as national and regional levels in relevant cases (Territorial Agenda 2011). Although the ideological imaginary of polycentric development reflects the multi-scalar nature of the territorial cohesion discourse, the operationalization of polycentricity differs at various spatial scales and in geographical contexts (Hall 2002), which contributes to multiple interpretations of polycentric development. Polycentricity is thus undoubtedly a scale-dependent concept (see Fig. 4.1). Thinking about scale is an absolutely crucial starting point for politically driven debates about polycentricity (see e.g. Taylor et al. 2008; Vasanen 2013). In terms of urban systems, the role of cities and towns is determined by the geographical level at which they act as centres (Malý and Mulíček 2016). Geographical scale may then completely change the intended effect of spatial planning tools. With increasing levels of mobility (human and non-human) and complexity of spatial and social networks, the issue of scale becomes even more important for urban planners. Besides, overlapping spatial scales are closely related to dynamic temporalities and the rhythms of specific linkages, processes, flows, events, and social actions (Marston and Smith 2001).

At the European level,[2] co-operation of the most developed cities and regions should create added value by contributing to the development of their wider regions (Territorial Agenda 2011). The support of polycentricity is focused on cities and regions that, through networking and linkages with other cities, can foster economic growth. The aim is to spread economic and socio-economically relevant functions among a larger number of urban centres in order to avoid further excessive economic and demographic concentration in the core area of the EU and ensure balanced development across the European territory (EC 1999). The existence of urban networks is expected, not only within EU borders (or Europe), but also beyond them, especially where linkages in cross-border territories and access to markets are an essential part of EU territorial development.

[2]Polycentricity could also, by operationalizing at the global level, point to the development of alternative global centres in terms of power and importance within the global economic network (London and partly Paris may be considered so-called 'global cities' within the European territory).

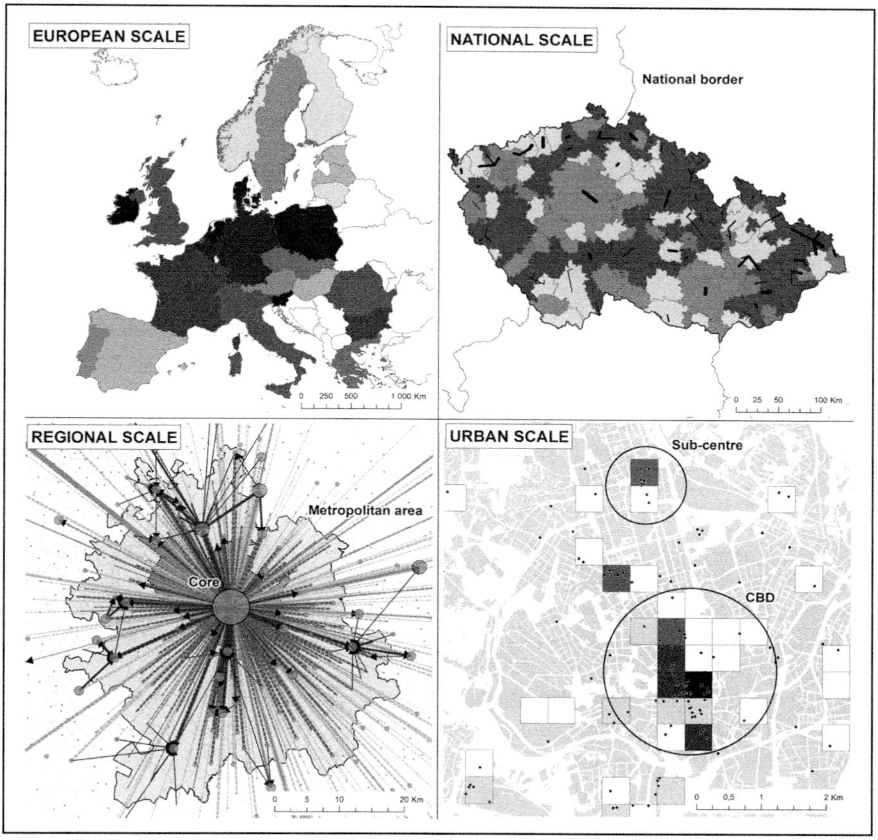

Fig. 4.1 Illustration of specific analytical treatment of polycentricity on different territorial scales (here specifically: European scale—index of polycentricity; national scale—cities rank-size distribution and significant commuting flows; regional scale—work-commuting patterns; urban scale—economic activities). *Source of data* ESPON (2005), CSO (2011), author's calculation

According to the ESDP, polycentricity at the European level means primarily promoting the economic prosperity of urban centres outside the so-called 'Pentagon' (the most economically advanced region in Europe, delimitated by London, Paris, Milan, Munich and Hamburg), i.e. mostly the capitals of other European countries or second-tier metropolitan areas lagging behind the largest metropolises (EC 1999). Although the EU is among the most economically advanced regions in the world, its internal structure is characterized by relatively large interregional disparities. The aim is to create more globally competitive and economically stronger areas that would ensure the future of the European economy (Davoudi 2003). Globally, an economically oriented approach is applied in territorial development strategies despite the presence of considerable environmental and social problems the economic core area of the EU is facing (Jensen and Richardson 2001).

Instead, the approach to territorial development, consisting of the utilization of the inner potential of individual regions, was represented earlier by a model called a 'Bunch of Grapes' (Kunzmann and Wegener 1991). Unlike the monocentric perception of Europe, it provides a polycentric vision, where the development of regions involves the exploitation of opportunities and advantages of each region (the placed-based rationale). In line with the notions of territorial and social cohesion, regions that are lagging behind should be given priority in terms of public funding. In a practical manner, on the one hand, polycentricity is, at the European level, understood as a principle ensuring competitiveness on the world market and, on the other hand, as a vehicle for remodelling existing regional disparities into a more territorially and socially cohesive form (Davoudi 2003).

At the nation-wide territorial scale, polycentric development is associated with balancing the dominant (usually the capital) city through the support of smaller regional urban centres. In this context, polycentric development should aim at balancing the spatial organization of human activities, which is expressed in the relationship between the core city and its hinterland within the urban region. So conceived, the interrelations between city regions of varying sizes and importance, and the interactions between urban regions with areas situated beyond their borders, should be taken into consideration when designing polycentric development strategies (Sýkora et al. 2009).

In view of this, the urban networks of regional centres have the potential to counterbalance the dominance of major cities by avoiding excessive territorial polarization between metropolitan areas and smaller regional centres and rural regions. These trends should help to reduce interregional disparities in the economic performance of individual regions. In the process, national strategic and spatial planning policies perceive polycentric development as a tool stimulating sustainability, integrated urban systems, and cohesion. As a general rule, partnerships among urban centres and countryside are highlighted, and special attention should be devoted to creating job opportunities and mitigating depopulation and socio-demographic deterioration trends in disadvantaged regions.

Territorial cohesion at the regional level should be reached by specific spatial patterns of daily social and economic interactions that boost regional economic performance whilst respecting environmental goals (Lambregts 2009). The concept of the polycentric urban region (PUR) was especially invented to define the structural characteristics of regional polycentricity and to provide an organizing framework for public-policy intervention (Kloosterman and Lambregts 2001). According to Parr (2004), the PUR is a distinctive form of region with potential for superior economic performance. Although the PUR represents a concept applied to subnational territories, it lacks a standard definition because of the spatial diversity of regions in terms of population density, the physical structure of urban systems, sociocultural traditions, the economic structure, and even the geographical scale (metropolitan polycentric systems and polycentric arrangements of small- and medium-sized towns can be distinguished; for example, the archetypal polycentric urban region Randstad in the Netherlands, characterized by a dense urban network with substantial economic performance, has different parameters when compared to the polycentric region of smaller towns situated in the countryside, yet they serve as centres for the region's rural hinterland).

Despite the somewhat ambiguous character of the PUR, the concept is mainly employed on the scale of daily urban systems, where the majority of the everyday commuting flows take place. The PUR is without a dominant centre and is composed of an integrated network of separate centres with a high intensity of mutual functional interactions. Specialization of centres, complementary assets, and a plurality of choices regarding jobs and public amenities are intrinsic aspects of the PUR. In contrast with monocentric urban systems, the PUR should avoid agglomeration diseconomies, high land prices, extensive territorial competition, infrastructure degradation, land fragmentation leading to urban sprawl and traffic congestion with all the negative side effects like pollution, noise, etc. (Bailey and Turok 2001). On the contrary, planning theory assumes that the PUR benefits from economic clustering, produces knowledge spillovers and generates a large supply of labour (Meijers 2008). From the viewpoint of governance, shared public policies and a multilevel institutional framework form the basis for efficient interventions and infrastructural projects, management of public spaces or commonly shared public amenities, and services of general interest.

The PUR is considered the most advanced conceptualization of a polycentric arrangement as several attempts to set down criteria defining the PUR have been recognized. As Parr (2004) suggests, a number of preconditions should be met in order to consider a region polycentric. From a morphological perspective, the PUR is constituted by a clustered distribution of centres where urban forms are clearly separated physically. The upper and lower limits of separation between centres are not precisely defined. However, the level of separation should reflect mobility habits, travel time, and average commuting distances. Furthermore, the PUR is defined by closer spacing of centres and a more even-size distribution of centres. As Mulíček and Malý (2018) claim, morphological spatial configuration is not a pre-given quality of a particular urban system. Rather it is an inherent part of the relational topologies of daily human lives and interactions, including work-commuting flows and other mobility-dependent activities with specific rhythms and routines (school, shopping trips, leisure activities, etc.).

From a functional perspective, a higher intensity of non-hierarchical economic interactions among the urban centres occurs in the PUR. Accordingly, the centres have different specializations and economic structures, and the market-oriented export base can be limited to the PURs inter-centre trade or distributed to (inter)national markets as well (Parr 2004).

When shifting the territorial scale to single urban forms, the municipal planning apparatus sometimes promotes the establishment of a polycentric city. Here, intra-urban sub-centres are usually determined by analysing urban spatial structures in terms of jobs, population distribution, and employment density. As reported by McMillen (2001), a polycentric city has one or more interacting employment sub-centre beyond the traditional central business district (CBD). Morphological urban structures have two dimensions that are commonly referred to in terms of mono/polycentricity patterns: centrality and clustering. Whilst the centrality dimension describes (de)centralized patterns of employment (or other variables) in relation to the CBD, the clustering dimension reflects the degree to which high-density subareas are clustered or randomly distributed (Tsai 2005; Pereira et al. 2013).

Sub-centres provide advantages for (not only suburban) workers and firms. The network of sub-centres is supposed to shorten commuting distances and costs and, consequently, have indirect impacts on the reduction of the traffic load, alleviation of excessive mobility, and in preventing city centre overcrowding. Furthermore, firms enjoy lower land costs and multiple choices for economic clustering. Urban planners and city decision-makers attempt to sustain the economic performance of local centres which serve as traditional places to offer jobs, services, and public amenities for surrounding neighbourhoods and are seen as a counterweight to the CBD.

Such local centres are often older towns which have been incorporated into expanding urban areas. On the other side, newly established nodes have begun to form around transportation crossroads situated further from the city centre (Anas et al. 1998). The city scale is thus confronted with hierarchical levels of the metropolitan region where widespread dispersion of economic activities is witnessed. In this context, polycentric city planning should bear in mind overlapping scales of the city and the higher order scale of surrounding urban forms which are closely tied to the urban core.

Promoting polycentricity at several territorial scales seems to be a contradictory step (Hall and Pain 2006). At the European level, polycentricity means strengthening the role of urban centres outside the most economically developed and densely populated area stretching from southeast England towards the south, across France and Benelux, to northern Italy. Thus, it primarily aims to support the capitals in Central and Eastern Europe whose economic growth potential emphasizes even more the monocentric structure of respective national urban systems (as in the case of Hungary, Czech Republic, Slovakia, and other countries with a major metropolis in a dominant position). In simple terms, the promotion of polycentric urban systems at the European level can lead to increasing monocentricity of urban systems at the national level. Analogically, a similar inconsistency can be observed on other territorial scales. It is, therefore, a fundamental question at which scales polycentric development should be planned to achieve its goal. Important too is to grasp what the implications are for other hierarchical levels of affected urban systems. When discussing polycentricity, urban planners, decision-makers, politicians, and other stakeholders need to realize the scale-dependency of polycentricity first. Otherwise, reaching polycentricity at the European level by concentrating economic activities into a few leading cities may cause regional inequalities between the core and periphery within each country (Hall and Pain 2006).

4.4 Normative Agenda Versus Factual Spatial Practices

Normatively defined, the spatial planning concept of polycentricity should contribute to lowering territorial disparities of the European territory and its regions. In spite of its conceptual ambiguity, polycentricity has quickly morphed into a policy buzzword, widely used within spatial planning practice and political debates. The concept of polycentric development is a magic wand that is supposed to cure all the socio-spatial troubles contemporary cities and regions are facing. But the crucial question posed is whether the support of polycentricity is based on empirical evidence, and if it is possible to prove that polycentricity has a positive impact on overcoming intra-regional disparities. Although the assessment of polycentricity and its association to spatial inequalities should be a fundamental part of the EU territorial cohesion-policy research agenda, only a few studies attempting to tackle this issue have been made so far. The main reasons for omitting this research stream are the ambiguity, multi-dimensionality, and multi-scalability of the polycentricity concept.

Despite these challenges, Meijers and Sandberg (2006) measured the extent of polycentricity in the national urban systems of 25 European countries and tested the hypothesis about the positive impact of polycentricity on regional disparities (GDP, unemployment rates). Their results point to the reverse logic: More monocentric urban systems are characterized by less regional disparities. Using morphological measures of polycentricity and socio-economic characteristics (GDP, regional employment, income distribution), Häzners and Jirgena (2013) similarly conclude that regional disparities in countries with polycentric urban system are not less pronounced than in countries with monocentric urban systems, and income distribution is not more equal in polycentric countries. Seen from another perspective, Veneri and Burgalassi (2012) assessed to what extent the degree of polycentric development is associated with various key indicators of economic, social, and environmental performance in the case of the Italian NUTS-2 regions. The findings are not in line with the normatively predefined assumptions about the positive impacts of polycentricity on regional growth, particularly in terms of social cohesion. In a different prism, Malý (2016a) investigated the relationship between the level of polycentricity (both in a morphological and functional sense) and the degree of intra-regional disparities (socio-demographic and economic indicators) by using functional coherent regions in Czech Republic as basic spatial units. The advantages of a polycentric urban arrangement were not confirmed in this case either. These rather sceptical conclusions of pioneering studies on the impact of polycentricity in overcoming territorial disparities, together with the still inadequately explored field of research, make the concept of polycentricity a topic that must be subjected to further and closer inspection.

Polycentric development, however, is not a static situation. Instead, it is a continually evolving and dynamic process. Hence, it is necessary to evaluate recent changes in spatial routines and the practices of contemporary society shaping the hierarchy of relational space in order to better understand the shifts of urban systems functioning in the context of the territorial cohesion discourse. In this framework, a recent study by Mulíček and Malý (2018) shows that there are discrepancies between specific representations of space stemming from the normative cohesion-policy agenda (polycentricity) and the spatial practices imprinted into the development and functioning of real urban systems. By raising a question to what extent spontaneously produced settings fit the imagination of polycentric and balanced normative space, they have been able to evaluate if the ongoing changes have been attaining a more polycentric organization of space, or not. By analysing changes in work-commuting flows (size and directions), the results show, however, that spatial patterns of urban systems constitute a path towards greater polarization, at least in the case of the Czech Republic, where the analyses were conducted. Hence, the shifts in the urban system hierarchy are far from the normatively defined balanced principles interweaving the territorial cohesion discourse.

Urban system evolution, conversely, is far more complex when urban systems continue to form multifaceted networks influenced by globalization, time-space compression, and metropolization processes, and cities are places experiencing growing consumption, mobility diversification, and social and household stratification. With reference to the new economic geography, polycentricity discourse has been enriched by concepts emphasizing the functional dynamics of urban systems and the spatial-economic position of particular towns and cities within an urban network (Partridge et al. 2009). In this context, the position of particular places and their ability to reap the benefits of economic networks become crucial aspects of these cities' development processes and strategies. Crucially, agglomeration theories are put into question by contradictory urban dynamics which have been seen in Western European cities, where the urban pattern of small- and medium-sized cities is a cornerstone of highly urbanized territories and where agglomeration externalities may be shared within city networks (Meijers et al. 2016).

Given this background, the concept of 'borrowed size' assumes that even smaller towns or cities are able to compensate for their lack of size by taking advantage of the proximity of a larger urbanized area and their embeddedness within a city network (Alonso 1973). In a broad assessment, small cities may borrow size from larger neighbours and exhibit the higher ranked characteristics of larger cities. Here, Meijers and Burger (2017) redefine the concept of 'borrowed size' by stretching it along several dimensions which address the scope (the difference between borrowing function and performance), the scale (different agglomeration externalities can be borrowed at various scales), and the need for further research revisiting the geographical foundations of agglomeration theory.

Some cities, however, may suffer from the proximity of a nearby metropolis in terms of less functions and performance produced (than expected given its size). By, experiencing this 'agglomeration shadow' effect, the growth of these cities is negatively influenced by higher tier urban centres in terms of exploiting their human capital and service functions (Meijers et al. 2016). A key question is: Which cities are prone to being covered by this 'agglomeration shadow'? And which tend to borrow the size of larger surrounding urban areas?

From a regional point of view, preliminary conclusions could be made regarding the provision of services and basic amenities of suburban localities. Municipalities strategically located on the axis of commercial zones and transport infrastructure as well as municipalities with the ability to gain benefits from a unique historical or cultural position are predominantly less affected by the shadow of agglomeration and borrow functions instead. Also, parts of municipalities in close proximity to a metropolis, and thus exposed to competition from larger urban centres, borrow performance in terms of population but demand higher ranked functions (Malý 2016b; Meijers et al. 2016).

A different methodical treatment is used for analysing polycentricity trends at (inter)national scales, where the association between the provision of high-end cultural amenities and integration into (inter)national urban networks has been evaluated. Here, the findings point to the occurrence of an 'agglomeration shadow' process which casts over smaller cities, and to the Christallerian logic that still dominates the urban system in North-West Europe (Burger et al. 2015). While research on the 'agglomeration shadow' and on the 'borrowed size' concepts continue, there is a strong need to reflect on such debates within the territorial cohesion discourse in order to redefine the interpretation of polycentric development. Altogether, particular attention should be paid to the underlying mechanisms standing behind the willingness of cities to co-operate and their ability to resist increased competition.

Although interpretation of the results of existing studies is limited due to methodological issues and data availability, the utmost attention should be given to general conclusions contesting the presumed effects of polycentric development. The key message here is that the concept of polycentricity needs a critical reflection as the alleged benefits of polycentric urban systems, in terms of diminishing spatial disparities, have not been fully proved. As Veneri and Burgalassi (2012) rightly point out, the evaluation of the normative principles of polycentricity, in relation to factual spatial practices, promise to be a stimulating field of research for the near future. Moreover, efforts to shed light on the still blurred concept of polycentricity should be made not only by incorporating the structural perspective of spatial and economic predispositions in urban systems but also by focusing on the institutional framework and soft spaces of co-operative relationships and their mechanisms, which are being established within structurally predefined polycentric urban systems.

4.5 Co-operative Relationships in Polycentric Urban Systems?

Pursuant to the EU territorial cohesion-policy agenda, spatial-functional configurations of polycentric urban systems represent structural preconditions for the improved socio-economic performance of the territories concerned. However, the empirical justification for the positive effects of polycentricity as it concerns balanced territorial development is missing, and the normative assumptions often linked to polycentric urban systems lack a solid analytical and conceptual treatment. In fact, the unfulfilled expectations of spatial planners and policymakers related to polycentric development are based especially on quantitative measurements of economic spatial structures and neglect for institutional frameworks of co-operation and coordination of regional development. In spite of several theoretical contributions on polycentric governance (McGinnis 1999; Ostrom 1999; Finka and Kluvánková 2015), efforts to interconnect structural and institutional perspectives of polycentricity, in order to analyse and assess the alleged benefits of polycentric urban systems, are almost absent.

Whilst the implementation of polycentric development strategies at the European and national territorial levels require coordination between spatial planners, sectoral policies, actors responsible for cross-border projects, and a reflection on the impacts of subsidy schemes in territories competing under market conditions, the achievement of the territorial cohesion objective, at the regional level, should be achieved through a bottom-up approach and co-operative relationships between local stakeholders. Surely co-operation among urban centres, which act as building blocks for 'a polycentric urban region', is a necessary element for effective public-policy interventions within the region concerned. Self-governing territorial units (municipalities) have to realize first, however, whether shared policies are a more efficient way to solve particular problems than the independent, but often limited policy agendas of individual cities and towns. However, even if a common policy is a solution, truly polycentric governance cannot be achieved without a proper institutional framework.

The issue of organizing the institutional framework of a polycentric urban system has been critically addressed by Meijers and Romein (2003), who suggest that governing a polycentric region requires new forms of regional coordination and the creation of the so-called 'regional organizing capacity'. In line with Albrechts (2001), they emphasize the need for co-operative regional arenas or forums which should be designed to support the process of building institutional or organizing capacity. Such a platform serves as a framework for discussing collectively shared problems, solutions, and planning policies among various kinds of actors (public actors, private companies, NGOs) as well as for initiating and implementing joint and coordinated actions. The potentialities of polycentric urban systems can be exploited only if new, even informal, networks of actors are established to bring up new ideas and projects of shared interest.

The expected benefits of inter-municipal co-operation, which only a polycentric urban region can provide, are reachable when a complex set of conditions is fulfilled. When perceiving inter-municipal co-operation as a bottom-up process, the building organizing capacity is based on social relations among actors with different needs and wants. Although social processes of negotiation and interaction take place within structurally non-hierarchical spatial arrangements, the form of collective action may be shaped unevenly with different expectations among particular actors about who is in charge and what is the balance of power. Informal coalitions may be a first step towards commonly managed projects. However, these should be followed by rather formal settings of collective action and rules framing mutual co-operation.

According to Meijers and Romein (2003), one of the reasons for the problematic implementation of organizing polycentric capacity is the persisting institutional fragmentation. In this sense, negotiating processes in a territory without rules of polycentric governance could be negatively affected simply due to the unintentional omission of some relevant actors or mutual misunderstandings stemming from different expectations. In the case of the Stockholm metropolitan region, Olsson and Cars (2011) point to the negligible impact of the regional polycentric strategy on local land-use strategies and national investments as a consequence of a missing binding authority responsible for polycentric governance.

In fact, individual municipalities often compete with each other by using rather extreme strategies, including rejection and opposition to a proposed development (NIMBY—'Not In My Back Yard') on the one side and the promising of unprofitable and unfeasible public projects on the other. At the municipal level, elected representatives and policymakers defend their own interests in order to fulfil promises they have given to voters during pre-election campaigns, which rarely include the 'greater good' of the higher ranked collective performance of the whole region. Such internal orientation of policymakers is one of the stumbling blocks in building a common regional identity (Meijers and Romein 2003) which is crucial for achieving shared objectives in a polycentric urban region (Franz and Hornych 2010; Olsson and Cars 2011).

Changing the traditional political mindsets of self-centred spatial development into the more contextual approach, by incorporating trans-scalar perspectives, seems to be one of the greatest challenges for urban planners and scientists with regard to realising polycentric spatial visions. Shaping new regional identities is a gradual process. Even if the decision-makers accept the polycentric urban region as a frame for territorial governance, the citizens may not share their vision. In line with Finka and Kluvánková (2015), academics and researchers should place the people's sense of territorial belonging and territorial responsibility at the forefront of polycentricity research.

From the spatial planning perspective, the regional scale of co-operation requires alternative planning approaches, including strategic spatial planning or structural plans (Albrechts 2001). While traditional land-use plans are more suited to municipal development, strategic spatial planning highlights the limited number of crucial strategic issues with particular emphasis on structural aspects (e.g. interconnecting centres in a PUR via a cycle route/path network) and the omission of details, non-significant at this stage and scale of planning (e.g. precise routing, construction materials). The next level of strategic planning approach includes action-oriented strategies and specific projects with a more detailed structure. Although this procedure is well known for urban strategies, applying it on a regional polycentric level involves new forms of co-operation, especially in terms of the type of actors who should participate. General agreement on crucial strategic issues may be accompanied by fierce negotiation and the occurrence of win–lose situations stemming from the conflicting interests of particular actors (Franz and Hornych 2010).

In spite of several obstacles being recognized for efficient inter-municipal co-operation in polycentric urban regions, there is growing evidence concerning the willingness of municipalities to co-operate on projects with a linear character. These projects promise major advances in the interconnection of municipalities and better accessibility of places that have, until now, been inadequately functionally linked to urban centres. Hence, the selection of crucial issues for polycentric urban systems include, predominantly, transport and traffic projects. For instance, from a long-term perspective, traffic and transport issues are the most threatening challenges for the Flemish Diamond (Albrechts 2001). Similarly, a co-operative arena for developing new transport infrastructure in the Stockholm region is supposed to strengthen its polycentric spatial structure. However, transport investments should be in line with land-use development strategies (Olsson and Cars 2011). In the same way, investigation of spatial structures in central Scotland brings evidence that only transport infrastructure policies partially exhibit features of a polycentric spatial planning strategy (Bailey and Turok 2001).

Transportation projects ensure the connection of places, link municipalities, and serve as a (physical) base for further co-operation. These are mostly projects that could not be financed and managed by a single municipal authority but require inter-municipal coordination. The development of transportation infrastructure may foster polycentric spatial patterns as well as monocentric arrangements of urban agglomerations where suburbs are better linked to the core. Co-operation on transportation projects is thus an effective tool to interconnect places, but it does not represent a specific approach used by polycentric urban regions with outcomes that would lead purely towards the establishment of polycentric structures. Consequently, the challenge is not to interconnect places by roads and paths, but to benefit from the physical interconnection of urban centres in the field of cultural and social coherence, environmental protection, and economic efficiency.

4.6 Conclusion

Polycentric development has gained a prominent position within urban and regional spatial planning policies since the concept holds great promise for bridging the cohesion/competitiveness dichotomy as a means to achieve territorial cohesion policy goals. While the structural perspective of polycentricity describes the internal spatial dynamics of a particular urban system, through a predefined pattern of physical spatial structure (morphological approach) and a relational space of inter-urban linkages (functional approach), the institutional perspective stresses the need for mobilizing inherent capacities, building organizing frameworks, networking different stakeholders, and incorporating action-oriented projects to gain benefits that a polycentric spatial structure may provide (Albrechts 2001). Regarding the EU cohesion-policy agenda, the 'polycentric obsession' continues despite the lack of empirical evidence about the positive impacts of polycentricity on harmonized balanced development. On the contrary, studies focusing on polycentric urban systems mostly end up with a sceptical conclusion about viewing polycentricity as a path to overcome regional imbalances.

Multiple geographical scales and territorial diversity stand behind the ambiguity of polycentricity when compared to the normative imagination of polycentric and balanced spaces (Mulíček and Malý 2018). In light of empirical studies gathered so far, the neoliberal political-economic model for promoting competitiveness, vis-a-vis the policy rationale for supporting more inclusive and cohesive policies, does not simply represent the natural paradigm contradiction which is a frequent topic in scholarly discussions. It also represents a tension transferred to the conceptualization of polycentricity, leading to conflicting interpretations of the alleged benefits of polycentric urban systems.

Enriching the territorial cohesion discourse, and the polycentricity debate in particular, with concepts and analytical treatments going beyond conventional agglomeration theories (i.e. confronting theories with discussion of urban dynamics, borrowed size, and agglomeration shadow) increases the chance of innovative conceptualization and operationalization of the polycentricity concept. The redefinition of polycentricity needs to question the expected ability of polycentric development to simultaneously generate growth and equilibrium.

The lack of functional, cultural, or political-institutional coherence and low identification with the region as a single spatial entity make smooth implementation of the polycentric governance processes difficult (Bailey and Turok 2001; Meijers and Romein 2003). While there is a willingness of municipalities to co-operate on transportation and traffic projects (and probably linear projects in general), most of the developmental issues that should be tackled efficiently at the regional level are approached from the level of self-oriented local governments because of the competitive mechanisms related to municipal rivalry and policy-driven motivations. Although empirical studies of co-operative relationships have been conducted for rather highly urbanized territories of first-ranked national urban centres, the research on the establishment of inter-municipal co-operation in polycentric urban regions should be amended to have a greater focus on polycentric spatial structures of smaller and medium-sized cities. This is where collective actions are of key importance as they substitute the benefits of agglomeration economies.

The ability to understand the interlinks between the 'grand narrative' of territorial cohesion, the planning imaginations of the European normative planning agenda, grounded territorial policies as well as factual urban systems structure and urban co-operation, belong to a pool of some of the greatest challenges confronting both urban planners and decision-makers in order to gain deeper insight into the bonds between polycentric spatial arrangements and solutions to specific problems particular territories are facing. The exaggerated belief in the polycentric spatial development as a panacea must be replaced by contextual thinking and the application of trans-scalar perspectives. Achieving territorial cohesion in terms of ensuring equal opportunities to all people, regardless of their place of residence or job location, requires strategic-oriented spatial planning. This includes site-specific solutions and drawing benefits from existing spatial-functional arrangements (even monocentric) as well as adaptation strategies of spatial development. In this regard, unrealistic planning visions of 'polycentric transformations' in traditionally monocentric structures with high economic performance should be modified into less normative strategies, including adaptive mobility mechanisms, alternative public policies (providing specific socio-economic programmes and projects), or support for acceptance of new territorial identities. Polycentric spatial development is thus one of the planning concepts which may or may not be used in reinforcing territorial cohesion of geographically and socio-economically diverse places.

References

Ache P, Andersen HT, Maloutas T, Raco M, Tasan-Kok T (eds) (2008) Cities between competitiveness and cohesion: discourses, realities and implementation. Springer, Berlin

Albrechts L (2001) How to proceed from image and discourse to action: as applied to the Flemish Diamond. Urban Stud 38(4):733–745

Alonso W (1973) Urban zero population growth. Daedalus 102:191–206

Anas A, Arnott R, Small KA (1998) Urban spatial structure. J Econ Lit 36(3):1426–1464

Bailey N, Turok I (2001) Central Scotland as a polycentric urban region: useful planning concept or Chimera? Urban Stud 38(4):697–715

Berry BJL (1964) Cities as systems within systems of cities. Pap Reg Sci Assoc 13(1):146–163

Böhme K, Gløersen E (2011) Territorial cohesion storylines: understanding a policy concept. Spatial Foresight Briefing 2011:1. Luxembourg

Burger M, Meijers E (2012) Form follows function? Linking morphological and functional polycentricity. Urban Stud 49(5):1127–1149. https://doi.org/10.1177/0042098011407095

Burger MJ, Meijers EJ, Hoogerbrugge MM, Tresserra JM (2015) Borrowed size, agglomeration shadows and cultural amenities in North-West Europe. Eur Plan Stud 23(6):1090–1109. https://doi.org/10.1080/09654313.2014.905002

Camagni R (1993) From city hierarchy to city network: reflections about an emerging paradigm. In: Lakshmanan TR, Nijkamp P (eds) Structure and change in the space economy. Springer, Berlin, pp 66–87

Christaller W (1933) Die Zentralen Orte in Süddeutschland. Gustav Fischer, Jena

CSO (2011) Census 2011. Available from: https://www.czso.cz/csu/sldb

Davoudi S (2003) Polycentricity in European spatial planning: from an analytical tool to a normative agenda. Eur Plan Stud 11(8):979–999. https://doi.org/10.1080/0965431032000146169

Davoudi S (2005) Understanding territorial cohesion. Plann Pract Res 20(4):433–441. https://doi.org/10.1080/02697450600767926

de Goei B, Burger MJ, van Oort FG, Kitson M (2010) Functional polycentrism and urban network development in the Greater South East, United Kingdom: evidence from commuting patterns, 1981–2001. Reg Stud 44(9):1149–1170. https://doi.org/10.1080/00343400903365102

EC (1997) Treaty of Amsterdam. Office for Official Publications of the European Communities, Luxembourg

EC (1999) ESDP—European spatial development perspective: towards balanced and sustainable development of the territory of the European Union. Office for Official Publications of the European Communities, Luxembourg

ESPON (2005) ESPON Project 1.1.1: potentials for polycentric development in Europe: project report. ESPON, Luxembourg

ESPON (2014) Making Europe open and polycentric: visions and scenarios for the European territory towards 2050. ESPON, Luxembourg

Evers D, Tennekes J, Borsboom J, Helingenberg H, Thissen M (2009) A territorial impact assessment of territorial cohesion for the Netherlands. Netherlands Environmental Assessment Agency (PBL), Hague

Faludi A (2004) Territorial cohesion: old (French) wine in new bottles? Urban Stud 41(7):1349–1365. https://doi.org/10.1080/0042098042000214833

Finka M, Kluvánková T (2015) Managing complexity of urban systems: a polycentric approach. Land Use Policy 42:602–608. https://doi.org/10.1016/j.landusepol.2014.09.016
Franz P, Hornych C (2010) Political institutionalisation and economic specialisation in polycentric metropolitan regions: the case of the East German 'saxony triangle'. Urban Stud 47(12):2665–2682. https://doi.org/10.1177/0042098009359951
Hall P (2002) Urban and regional planning. Routledge, London
Hall P, Pain K (eds) (2006) The polycentric metropolis: learning from mega-city regions in Europe. Earthscan, London
Häzners J, Jirgena H (2013) Polycentricity measures and regional disparities. In: Annual 19th international scientific conference proceedings: research for rural development 2013, May 15–17, Latvia University of Agriculture, Jelgava
Jensen OB, Richardson T (2001) Nested vision: new rationalities of space in European spatial planning. Reg Stud 35(8):703–717. https://doi.org/10.1080/00343400120084696
Kloosterman RC, Lambregts B (2001) Clustering of economic activities in polycentric urban regions: the case of the Randstad. Urban Studies 38(4):717–732. https://doi.org/10.1080/00420980120035303
Kloosterman RC, Musterd S (2001) The polycentric urban region: towards a research agenda. Urban Stud 38(4):623–633. https://doi.org/10.1080/00420980120035259
Kunzmann KR, Wegener M (1991) The pattern of urbanisation in Western Europe 1960–90. Institut für Raumplanung, Universität Dortmund, Dortmund
Lambregts B (2009) The polycentric metropolis unpacked: concepts, trends, and policy in the Randstad Holland. Doctoral dissertation, Amsterdam Institute for Metropolitan and International Development Studies, Amsterdam
Malý J (2016a) Impact of polycentric urban systems on intra-regional disparities: a micro-regional approach. Eur Plan Stud 24(1):116–138. https://doi.org/10.1080/09654313.2015.1054792
Malý J (2016b) Small towns in the context of "borrowed size" and "agglomeration shadow" debates: the case of the South Moravian region (Czech Republic). Eur Countrys 8(4):333–350. https://doi.org/10.1515/euco-2016-0024
Malý J (2018) Questioning territorial cohesion: (un)equal access to services of general interest. Pap Reg Sci 97(2):323–343. https://doi.org/10.1111/pirs.12250
Malý J, Mulíček O (2016) European territorial cohesion policies: parallels to socialist central planning? Moravian Geogr Rep 24(1):14–26. https://doi.org/10.1515/mgr-2016-0002
Marston SA, Smith N (2001) States, scales and households: limits to scale thinking? A response to Brenner. Prog Hum Geogr 25(4):615–619. https://doi.org/10.1191/030913201682688968
McGinnis MD (ed) (1999) Polycentric governance and development: readings from the workshop in political theory and policy analysis. University of Michigan Press, Ann Arbor
McMillen DP (2001) Nonparametric employment subcenter identification. J Urban Econ 50(3):448–473. https://doi.org/10.1006/juec.2001.2228
Medeiros E (2016) Territorial cohesion: an EU concept. Eur J Spatial Dev 60. Available from: http://www.nordregio.org/publications/territorial-cohesion-an-eu-concept/
Meijers E (2008) Summing small cities does not make a large city: polycentric urban regions and the provision of cultural, leisure and sports amenities. Urban Stud 45(11):2323–2342. https://doi.org/10.1177/0042098008095870
Meijers EJ, Burger MJ (2017) Stretching the concept of 'borrowed size'. Urban Stud 54(1):269–291. https://doi.org/10.1177/0042098015597642
Meijers E, Romein A (2003) Realizing potential: building regional organizing capacity in polycentric urban regions. Eur Urban Reg Stud 10(2):173–186. https://doi.org/10.1177/0969776403010002005
Meijers E, Sandberg K (2006) Polycentric development to combat regional disparities? The relation between polycentricity and regional disparities in European countries. In: Proceedings of the 46th congress of the European Regional Science Association, August 30–September 3, Volos, ERSA

Meijers EJ, Burger MJ, Hoogerbrugge MM (2016) Borrowing size in networks of cities: city size, network connectivity and metropolitan functions in Europe. Pap Reg Sci 95(1):181–198. https://doi.org/10.1111/pirs.12181

Mulíček O, Malý J (2018) Moving towards more cohesive and polycentric spatial patterns? Evidence from the Czech Republic. Pap Reg Sci. Early view online, https://doi.org/10.1111/pirs.12383

Olsson AR, Cars G (2011) Polycentric spatial development: institutional challenges to intermunicipal cooperation. Jahrbuch für Regionalwissenschaft 31(2):155–171. https://doi.org/10.1007/s10037-011-0054-x

Ostrom E (1999) Coping with tragedies of the commons. Annu Rev Polit Sci 2:493–535. https://doi.org/10.1146/annurev.polisci.2.1.493

Ostrom V, Tiebout C, Warren R (1961) The organization of government in metropolitan areas. Am Polit Sci Rev 55:831–842

Parr JB (2004) The polycentric urban region: a closer inspection. Reg Stud 38(3):231–240. https://doi.org/10.1080/0034340042000211114

Partridge MD, Rickman DS, Ali K, Olfert MR (2009) Do new economic geography agglomeration shadows underlie current population dynamics across the urban hierarchy? Pap Reg Sci 88(2):445–466. https://doi.org/10.1111/j.1435-5957.2008.00211.x

Pereira RHM, Nadalin V, Monasterio L, Albuquerque PHM (2013) Urban centrality: a simple index. Geogr Anal 45(1):77–89. https://doi.org/10.1111/gean.12002

Robert J (2007) The origins of territorial cohesion and the vagaries of its trajectory. In: Faludi A (ed) Territorial cohesion and the European model of society. Lincoln Institute of Land Policy, Cambridge, MA, pp 23–35

Sýkora L, Mulíček O, Maier K (2009) City regions and polycentric territorial development: concepts and practice. Urban Res Pract 2(3):233–239. https://doi.org/10.1080/17535060903319095

Taylor PJ, Evans DM, Pain K (2008) Application of the interlocking network model to mega-city regions: measuring polycentricity within and beyond city-regions. Reg Stud 42(8):1079–1093. https://doi.org/10.1080/00343400701874214

Territorial Agenda (2007) Territorial agenda of the European Union. Agreed on 24/25 May 2007, Commission of the European Communities, Leipzig

Territorial Agenda (2011) Territorial agenda of the European Union 2020. Agreed on 19 May 2011, Commission of the European Communities, Gödöllő

Tsai Y-H (2005) Quantifying urban form: compactness versus 'Sprawl'. Urban Stud 42(1):141–161. https://doi.org/10.1080/0042098042000309748

Vanolo A (2010) European spatial planning between competitiveness and territorial cohesion: shadows of neo-liberalism. Eur Plan Stud 18(8):1301–1315. https://doi.org/10.1080/09654311003791358

Vasanen A (2013) Spatial integration and functional balance in polycentric urban systems: a multi-scalar approach. Tijdschrift voor Economische en Sociale Geografie 104(4):410–425. https://doi.org/10.1111/tesg.12029

Veneri P, Burgalassi D (2012) Questioning polycentric development and its effects: issues of definition and measurement for the Italian NUTS 2 Regions. Eur Plan Stud 20(6):1017–1037. https://doi.org/10.1080/09654313.2012.673566

Waterhout B (2007) Territorial cohesion: the underlying discourses. In: Faludi A (ed) Territorial cohesion and the European model of society. Lincoln Institute of Land Policy, Cambridge, MA, pp 37–59

Waterhout B (2008) The institutionalisation of European spatial planning. IOS Press, Amsterdam

Jiří Malý is a researcher at the Czech Academy of Sciences, Institute of Geonics and a specialist in urban geography at the City Chief Architect's Office in Brno. He completed his Ph.D. in Regional Geography and Regional Development at Masaryk University (Brno) with special focus on the concept of polycentricity as a part of territorial cohesion discourse. His research interests lie predominantly within urban and regional development, urban systems, spatial planning, metropolitan processes and urban change. He has been involved in working groups processing spatial planning documents and strategic visions for urban development and in research projects focused on regeneration of urban environment, dynamization of spatial data or urban mobility.

Chapter 5
Cross-Border Functional Regions and Innovation: Cases from Inner Scandinavian Regions

Bjørnar Sæther, Bjørn Terje Andersen and Erik Hagen

Abstract This chapter describes and analyses the evolution of three cases of cross-border innovation systems in Inner Scandinavia. Functional interdependence between partners on either side of the Sweden–Norway border has emerged. Interreg Inner Scandinavia has long been instrumental in these developments, coordinating planning and providing support for innovations in the region. Politicians, municipal officers, business representatives and experts have all acted as entrepreneurs. Over time a new path within music management has been created, as Inner Scandinavia has become an important node in the global production of commercial music. Firms and knowledge providers within clean technology have attained key positions from their bases in Inner Scandinavia. The SITE border region is central to Scandinavia's winter tourism industry. This demonstrates the possibility of establishing new knowledge-based economic activities in the rural borderlands between Sweden and Norway.

Keywords Cross-Border innovation · Functionality · Inner Scandinavia
Rural development · Tourism · Music management

5.1 Introduction

This chapter investigates learning and innovation as part of cross-border functionality in rural areas in Inner Scandinavia. Functionality will be discussed with reference to definitions of working and living regions according to Statistics Norway. Definitions

B. Sæther (✉)
Department of Sociology and Human Geography, University of Oslo, Oslo, Norway
e-mail: bjornar.sather@sosgeo.uio.no

B. T. Andersen · E. Hagen
Hedmark County Council, Hamar, Norway
e-mail: bjoern-terje.andersen@hedmark.org

E. Hagen
e-mail: erik.hagen@hedmark.org

E. Medeiros (ed.), *Territorial Cohesion*, The Urban Book Series,
https://doi.org/10.1007/978-3-030-03386-6_5

are largely based on commuting distance and hence their utility is limited when investigating the role of learning and innovation in strengthening functional regions.

The concept of functional distance refers to differences between regions in innovation performance (Lundquist and Trippl 2013; Maggioni and Uberti 2007). Functional distance serves as an analytic point of departure for the discussion of cross-border regional innovation systems. Three ideal types of cross-border innovation systems have been proposed: weakly, semi- and strongly integrated systems (Lundquist and Trippl 2013). The relevance of the strongly integrated ideal type in a rural context will be discussed. Public policy and in particular the role of Interreg programs will be discussed as well.

Subsequently, three cases of cross-border integration of learning and innovation will be presented. The first case discusses cross-border cooperation in tourism and destination development in four border municipalities in the Dalarna-Hedmark (in Sweden and Norway, respectively) region. The second case discusses a music ecosystem in which the cities of Arvika and Karlstad in Sweden and Rena in Norway represent key nodes. Cross-border cooperation in clean technology is the third case. The cases will be compared and the degree to which they are aspects of cross-border innovation systems will be examined. Any theoretical lessons that can be learnt concerning cross-border innovations systems in rural regions will also be discussed. Furthermore, some adjustments of the ideal types of cross-border innovation systems proposed by Lundquist and Trippl (2013) will be proposed.

Our primary research question is thus the following:

- How can cooperation in innovation help strengthen cross-border functionality in a rural region?

Two more specific research questions will also be discussed:

- What is the role of Interreg in supporting cross-border learning and innovation?
- What role is played by system-level entrepreneurs?

5.2 Methodology

This is a theoretically informed case study based on a qualitative intensive investigation of three cases. As such, it is part of an important methodological tradition in human geography and related disciplines (Danermark et al. 2002; Ragin and Amoroso 2011). Sources such as interviews with stakeholders, written sources regarding cross-border cooperation from the past 20 years and published academic work on cross-border cooperation in Scandinavia and elsewhere were triangulated for this purpose.

Two of the authors have long served as public servants in Hedmark County Council (Norway), and have played a central role in the policy formulation and implementation of the Interreg programmes. This experience has provided them with profound insights into the development of cross-border cooperation on the ground as well as the role of Interreg programmes in supporting such cooperation over the past 20 years. The insider role of the two authors can be connected with the theoretical and methodological competencies of the third author, who is an outsider.

The writing of this chapter thus resembles the final step in a process of co-generation of knowledge as formulated by Karlsen and Larrea (2014). A process of participatory change and cogenerated learning in territorial development consists of four steps: problem definition, communication and learning, problem solving and learning from practice. This indicates that insiders and outsiders work together in all phases to solve practical problems of regional development. However, in the writing of this chapter, this dynamic was not apparent. The insiders worked with the outsider only in the final (ex post) phase of learning from practice. The study is, therefore, traditional in its reporting ex post, but it also contains an element of cogenerated learning between insiders and an outsider.

5.3 Functionality and Innovation

The categorisation of functional regions represents an important form of nomenclature in dividing territories into geographical units. The dominant way of defining functional regions is in terms of a town or city and its surroundings. In Norway, there are 160 such functional working and living regions (Gundersen and Jukvam 2013). Travel distance, normally no more than 75 min by car, and share of people commuting are among the criteria presented by Gundersen and Jukvam (2013). This understanding of functionality is central to urban and regional planning, policy making and statistics. In a context of cross-border cooperation, the question of whether municipalities along the Swedish–Norwegian border are part of functional regions has received attention. Based on data from 2009, the share of cross-border commuting was too low to identify the existence of any cross-border functional region (Ørbeck and Braunerhielm 2013). The Swedish municipality with the highest level of commuting to Norway was Årjäng, where 18% of the workforce commuted in such a way. However, fewer than 10% commuted to a particular functional region on the Norwegian side of the border. The number of Swedish commuters to Norway is too small and not sufficiently concentrated for the municipalities on either side of the border to be deemed functionally integrated working and living regions. The number of commuters from Norway to Sweden is even smaller. According to the definition of functionally integrated working and living regions proposed by Gundersen and Jukvam (2013) and accepted by Statistics Norway, there are no cross-border functionally integrated regions here.

This prompts us to ask whether a definition of functional integration be limited to commuting and travel times. It has been argued that '[t]he idea of the functional region captures the idea of a territory characterised by spatially related human activities' (Tomaney 2009, 140). Spatially related activities include much more than commuting. In a proposed EU regulation, the importance of supporting functional cross-border regions has been emphasised. Based on evaluations of Interreg programmes, it was noted that '[l]imited attention seems to have been paid to the notion of a functional region or area when identifying the regions to support. However, this is essential to considering the potential benefits of cross-border transnational cooperation' (European Commission 2018, 5).

This would suggest that the European Commission has a broad understanding of functional regions based on cooperation, something that is underlined by its proposal for the next Interreg period, 2021–2027. This states that cooperation will be further strengthened by '[a]dapting the architecture of Interreg programmes to take better account of functional areas. Cross-border programs will be better streamlined in order to concentrate resources on land borders where there is a high degree of cross-border interaction' (European Commission 2018, 5). This implies that discussions of territorial functionality are rising to the top of the policy agenda within the EU.

The concept of functional distance was introduced by Maggioni and Uberti (2007) to refer to differences between regions in innovation performance. Based on these authors' work, Lundquist and Trippl (2013, 453) argue that 'knowledge does not flow easily between areas if they differ strongly in their innovation capacity'. This notion of functional distance and what it means for knowledge flows can be placed within a larger debate within economic geography concerning proximity (Boschma 2005). A fine balance exists between partners being sufficiently close for knowledge exchange and yet far enough away to learn something they did not already know. Maggioni and Uberti conceptualise functional distance as the difference in absolute registered values of two regions on the European Commission's Regional Innovation Scoreboard. Key findings from the scoreboard for North Middle Sweden and Hedmark-Oppland in Norway are listed in Table 5.1.

It should be noted that the data were calculated at the Nomenclature of Territorial Units for Statistics (NUTS) 2 level, which includes two counties that are not within Inner Scandinavia: Gävleborg County in Sweden and Oppland County in Norway. According to the scoreboard, the regions are strong innovators and interestingly share many of the same strengths and weaknesses regardless of whether they are situated in Sweden or Norway. Innovative small and medium-sized enterprises (SMEs) in both

Table 5.1 Regional Innovation Scoreboard

	Overall performance	Strengths	Weaknesses
North Middle Sweden, consisting of the counties of Värmland, Dalarna and Gävleborg	Strong innovator; performance increased over time	Innovative SMEs collaborating; lifelong learning	Public–private co-publication in research
The counties of Hedmark and Oppland in eastern Norway	Strong innovator with performance increasing over time	Innovative SMEs collaborating; lifelong learning	Public–private co-publication; R&D spending on businesses

Source http://ec.europa.eu/growth/industry/innovation/facts-figures/regional_en. Downloaded 20 June 2018

regions are considered strong in terms of collaboration. The fact that innovative SMEs collaborate in rural areas was previously reported by Jakobsen and Lorentzen (2015); indeed, SMEs in rural areas tend to cooperate more than their urban counterparts. Collaboration in rural areas can be deemed a strategy to compensate for long distances and less institutional support for innovation. Both regions also score highly in terms of lifelong learning; in fact, North Middle Sweden is rated among the European leaders in this respect. The regions additionally share a number of structural characteristics. For instance, gross domestic product (GDP) per capita is almost identical between the two regions, and considerably lower than the national average, especially in Norway. Both regions have a higher share of employment within manufacturing, measuring 15.5% on the Swedish side and 9.9% on the Norwegian side. It can, therefore, be concluded that the functional distance between the two regions is low, facilitating cross-border knowledge flows. However, a low functional distance does not by itself guarantee cross-border learning and innovation. Empirical studies are necessary to investigate whether and how learning and innovation take place.

Knowledge flows between firms and knowledge producers are central to the regional innovation system (RIS) analytic approach, which has emerged during the past 25 years (Asheim and Gertler 2005). The RIS approach is a systemic approach to innovation that consists of two subsystems: knowledge producers and knowledge users. Its systemic approach intimates less interest in individual actors within the RIS. In order to more effectively analyse the three cases in this study, some of the key actors will be addressed.

Inner Scandinavia is a predominantly rural region, but it is located on the Oslo—Stockholm axis, the latter of which being (the) most innovative region in Europe according to the European scoreboard. However, owing to shorter distances and considerable growth in the Oslo region, Inner Scandinavia is in some respects more strongly related to Oslo than to Stockholm. This is supported by Region Värmland's slogan 'Värmland makes the Oslo region larger'. Region Värmland has formulated its own Norway strategy in order that it might benefit from its relative proximity to Oslo. Värmland is thus following a strategy of linking up with stronger regions, as recommended as a general strategy (Tödtling and Trippl 2005). Being connected with a more innovative region represents one means by which rural regions compensate for a lack of innovative capacity. Rural regions are often dominated by SMEs operating in traditional or resource-based industries and lack what has been termed 'related variety'. In less variegated and less industrially dense regions, fewer possibilities exist to combine different types of knowledge (Isaksen and Karlsen 2016).

In discussing regional industrial development in regions within RIS, Rypestøl (2018) argues that path extension represents the most plausible way of advancing. In such regions, external ideas are important and can be combined with pre-existing local knowledge (Rypestøl 2018, 55). Given that most firms are small and/or lack the capacity to work systematically with research and innovation, system-level entrepreneurs play a potentially important role. System-level entrepreneurs can work to bridge gaps between firms, between firms and knowledge providers and between firms and (local) government officials. Such efforts often require considerable time, but they can eventually create extensive networks. In order to achieve these goals, system-level entrepreneurs depend upon long-term financing. This can be provided by programmes supporting regional development and innovation. As will be described below, Interreg programmes in Inner Scandinavia have supported system-level entrepreneurs. Figure 5.1 presents a typology of cross-border regional innovation systems as proposed by Lundquist and Trippl (2013).

This typology summarises the important characteristics and challenges connected to cross-border cooperation in innovation. As a typology, it is general in character and the discussion here will focus on the ways in which it can enable analysis of cross-border development in rural areas. The Stage III, most developed form of cross-border RIS is of limited importance in rural areas. Stage III is a symmetrical, innovation-driven system with related variety and functional proximity in a wide range of business areas, according to Lundquist and Trippl (2013). However, rural areas tend to be deficient in these regards. Indeed, if a cross-border innovation system were to fulfil all of the criteria of a fully integrated cross-border innovation system, it would probably not be located in a rural area.

However, the other two typologies are useful for analysing cross-border learning and innovation in rural areas. In Stage II, the economic structure is characterised by emerging synergies and functional proximity in a few business areas, while in Stage I this is almost entirely lacking. Mechanisms for the coordination of innovation policies emerge in Stage II (Lundquist and Trippl 2013). The typology also includes as its dimensions knowledge infrastructure, accessibility and institutional setup. This renders Stages I and II of the typology useful for an analysis of learning and innovation in Inner Scandinavia.

Public policy and policy leadership can potentially play an important role in supporting innovation and SMEs in rural areas. The role of public policy in such

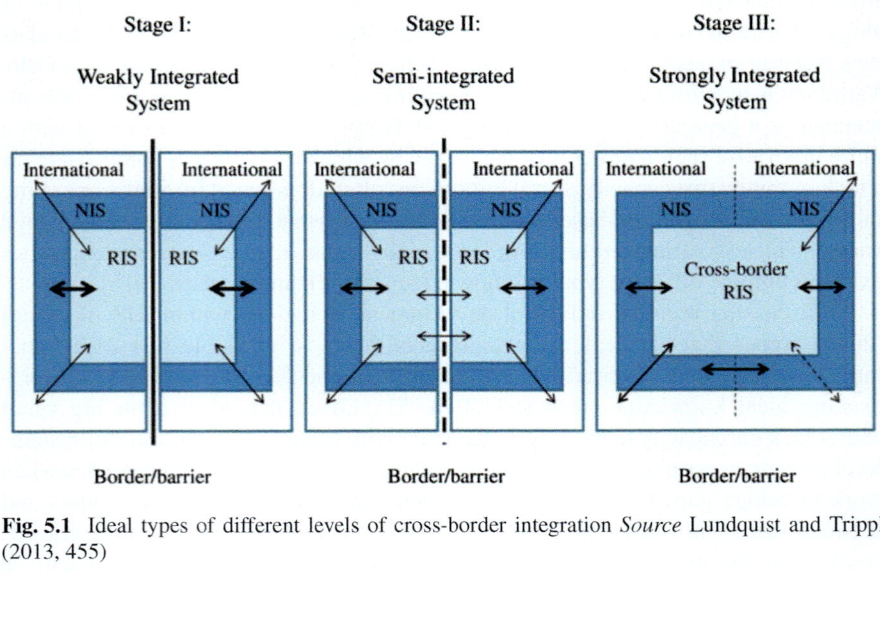

Fig. 5.1 Ideal types of different levels of cross-border integration *Source* Lundquist and Trippl (2013, 455)

locations is qualitatively different from urban areas and economically and technologically dynamic regions (Tödtling and Trippl 2005). In the overview of different stages of cross-border RIS, policy structures are characterised by the three stages. In Stage I, when systems are poorly integrated, an absence of policy leadership with any vision or legitimacy can be noted. Scant or asymmetrical support from the nation-states is prevalent. In Stage II, mechanisms for the coordination of innovation policies begin to emerge. In strongly integrated systems in Stage III, transparent and democratic governance structures are in place, with inclusive forms of governance and civic participation (Lundquist and Trippl 2013).

So far, we have presented some of the existing literature on functionality and innovation concerning cross-border regions. This literature will guide the analysis of cases from Inner Scandinavia.

5.4 Cross-Border Cooperation in Inner Scandinavia

Inner Scandinavia (Fig. 5.2) is a large geographical area with a low population density. It is a heterogeneous region and comprises large areas mainly located to the north of the Oslo–Stockholm axis, consisting of the counties of Hedmark and Akershus in Norway and Värmland and Dalarna in Sweden. Some of the municipalities to the north in Hedmark, Värmland and Dalarna are experiencing absolute population decline, loss of working places and processes of marginalisation.

Cross-border cooperation between Norway and Sweden has a long history. Starting in the 1960s, the Nordic Council encouraged Nordic cooperation and helped formalise this by establishing border committees. At the regional level, Hedmark County Council and Länstinget in Värmland adopted a political initiative to support cross-border cooperation in the fields of culture, infrastructure, innovation and competence. At that time, neither Sweden nor Norway was a member of the European Economic Community (EEC).

Sweden became an EU member in 1994, whereas Norway opted against joining and instead became affiliated with the EU through the European Free Trade Association (EFTA) and the European Economic Area (EEA) Agreement. Sweden's EU membership marked the introduction of new mechanisms in Swedish–Norwegian cooperation. Sweden became part of the EU Cohesion Policy, including European territorial cooperation. The Norwegian government and Norwegian border regions have since 1996 been invited to participate in cross-border cooperation (CBC) programmes by the Swedish government. Such a contribution from the Norwegian government is provided via the national budget as well as co-financing from partnerships at the regional level.

The EU Interreg programmes have supported cross-border cooperation in Inner Scandinavia since the 1990s, from the Interreg II-A programme 1996–1999 until the present Interreg V-A 2014–2020 (Medeiros 2014a). From the outset, cross-border cooperation aimed to support cooperation between neighbouring administrative authorities located close to internal or external EU borders (Medeiros 2014b).

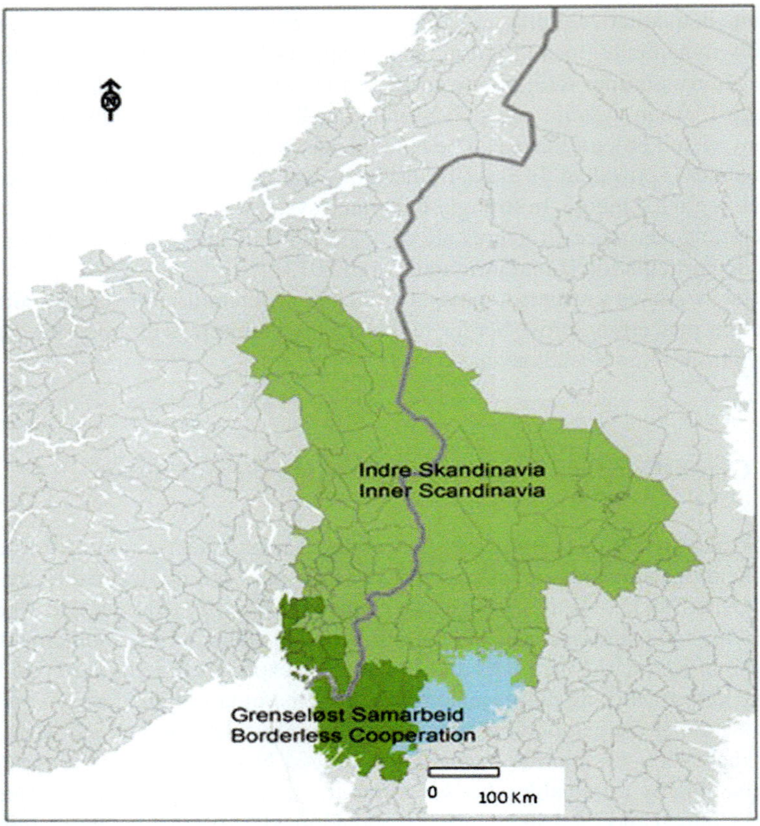

Fig. 5.2 Inner Scandinavia *Source* Own elaboration

Over the years, the content of the programmes has changed. The importance of developing economic sectors or 'clusters' of firms has increased, alongside support in competence, research and development. At the start of the latest programme period, 2014–16, SMEs represented the main beneficiaries, together with research and development receiving 80% of the available funding. This indicates that a systemic approach towards SME development and innovation has achieved significance within Interreg.

Four axes of intervention in the programming phases of Sweden–Norway cooperation within Interreg 1996–2016 can be identified (Medeiros 2017). In the first period, 1996–1999, the four axes were: (1) cultural identity; (2) information, services and transport; (3) business and sector development; and (4) skills and competence. Between 20 and 30% of the total funds were allocated to each of the four axes. Conjointly, 60% of total funds were allocated to business and sector development and skills and competence. Regarding Interreg V and the first two years 2014–2016, as

Table 5.2 Indicators of research and innovation

Indicators of priority research and innovation	2023 result for Inner Scandinavia	Anticipated numbers as of 31 December 2016[a] in Inner Scandinavia
Number of SMEs participating in cross-border research, development and innovation activities	50	233
Number of research institutes participating in cross-border research projects	5	27
Demonstration and test projects	2	17
Number of long-term formalised cooperation agreements between research institutes	2	9

Source Alnes and Sæther (2017)
[a]Source: Interreg Projects Progress Reports 2017
The indicators demonstrate that a much greater number of SMEs than anticipated are participating and that the programme-level targets have already been met

previously mentioned, a large share of total funds was allocated to research and development and SMEs. Thus, funding for research and SME development has always been of considerable importance to the Sweden–Norway Interreg programmes.

The systemic dimension of support for SMEs has increasingly been at the forefront of the Interreg programmes' agendas, and economic support is not provided to single firms. One of the projects currently being supported is the Music Ecosystem (MECO) in Inner Scandinavia. It has been argued that '[i]nnovation systems are not primarily characterised by competitive strength created in single firms, but within business and industrial environments' (authors' translation).

A priority in the current programme period is to advance the research, development and innovation capacity of organisations and businesses. This priority attracts SMEs, with examples of relevant initiatives being projects that increase cooperation and contribute to knowledge transfers between research institutions, businesses/SMEs, the public sector and civil society, as well as projects that utilise a smart specialisation approach to develop the strengths of the cross-border region. The indicators below (Table 5.2) demonstrate the programme-level targets and the progress that has been made to achieve them as of 31 December 2016.

5.5 Cases of Emerging Cross-Border Innovation

5.5.1 Cross-Border Tourist Destination Development

At the heart of Inner Scandinavia, the four municipalities of Malung-Sälen and Älvdalen in Dalarna County on the Swedish side and Trysil and Engerdal in Hedmark County in Norway cooperate in the Sälen Idre Trysil Engerdal (SITE) region. The region is mountainous and experienced many years of population decline owing to the loss of employment in agriculture and forestry. However, from the 1960s, local politicians and entrepreneurs began to develop winter tourism activities, establishing the region's first ski lift. The long-time mayor of Trysil was instrumental in this development. From the 1970s, Trysil emerged as one of the most popular winter destinations in Norway, and people from the Oslo/Akershus region started to construct their own cabins here. On the Swedish side, Sälen experienced a parallel development. These municipalities were early in establishing themselves as their nations' leading winter tourism destinations (Table 5.3).

In contrast to the situation in the southern part of Inner Scandinavia, a long tradition of cross-border cooperation between Dalarna and Hedmark did not exist (Nauwelaers et al. 2013). However, from the mid-2000s, Interreg Inner Scandinavia financed projects GREEN 1, 2 and Green 2020 in the SITE region, whereby municipalities and local firms joined forces to develop more environmentally friendly energy production and consumption. These projects were organised as private–public partnerships. Based on a series of practical and often cost-saving results in reducing energy consumption and strengthening local bio-based energy production, firms and municipalities learnt to appreciate cooperation.

All actors found common ground in striving for a shared environmentally friendly destination label as a means of boosting the market competitiveness of the destination or SITE region as a whole. They agreed to work to develop the SITE region into a world-class and cross-border tourist destination. The Interreg Inner Scandinavia programme has been a key partner in supporting the four border municipalities in working towards this goal through the SITE 1 and 2 projects, which has spanned for nine years from 2010 to 2018. The project has facilitated the influx of external resources that have helped make the ambition more specific and knowledge based.

The municipalities, including their politicians and managers, work together. A specific concern is that the future development of the region should be knowledge based. Experts and researchers are utilised as advisors for the region's branding and planning issues. Local firms in the construction industry or those that offer experience-based tourism activities are included in the project work. Local farmers have started to offer high-quality food to tourists. Managers in the municipalities are highly appreciative of this historic cooperation. As an effect of the SITE cooperation, local officials have come to perceive their municipalities as more closely related to the neighbouring SITE municipalities than to other municipalities in their own county. This is due to functional similarities, namely the dominant position of a highly specialised and globally oriented tourism sector. Through the SITE cooperation,

Table 5.3 Annual turnover in tourism in the SITE region, 2016

Sälen	Idre/Älvdalen	Trysil/Engerdal	SITE-region
250 mill EUR	70 mill EUR	120 mill EUR	440 mill EUR

Source SITE project plan and Trysil municipality

municipalities and firms also work together to solve typical border issues concerning customs, transport and welfare provision.

Key actors in the region have long lobbied for the construction of an airport to transport tourists to the region. In 2017, a decision was made to construct the Scandinavian Mountains Airport, also known as Sälen-Trysil Airport, which is scheduled to open in 2019.

SkiStar is the major private firm in the region and operates ski resorts on both sides of the border. SkiStar runs lifts, hills, lodging, restaurants and other services in Trysil and Sälen. In the financial year 2016–2017, SkiStar reported a turnover of €33 million in Sälen and €25 million in Trysil. Indeed, the firm's profitability is good. According to the CEO, the firm expects to see a considerable growth in tourism in the region. SkiStar's vision is to 'create memorable mountain experiences as the leading operator of European alpine destinations'. SkiStar is listed on the Stockholm stock exchange and a majority of the stocks are owned by brothers Mats and Erik Paulsson, who are among Sweden's financial elite and live in the southern part of the country. This means that important decisions concerning the future development of tourism in the SITE region are made by this family and the SkiStar board.

Winter tourism as a new path to economic development has, in many ways, created in the SITE region. Cooperation between the municipalities and between private and public actors, together with a history of political, social and political entrepreneurship, has explained much of its progress. In collaboration, local actors have become sufficiently strong to push for the construction of an airport in the region.

5.5.2 *Knowledge Production in the Music Industry*

In the past 15 years, the town of Rena in rural Hedmark County has emerged as a national leader in knowledge production within music management and production. Moreover, Rena has become a local node in the global knowledge support system for commercial music. This fascinating development would not have been possible without the support of Interreg in Inner Scandinavia and cooperation with Swedish partners.

In the period 2003–2008, when Internet-based innovations destroyed the traditional music industry, entrepreneurs at Rena started to educate the next generation of music managers. Whereas managers in the traditional music industry were sceptical of its future, young students had a vision. Their thinking was based on the simple fact that although people were increasingly unwilling to pay for music, they continued to listen to it and there was no sign that this would stop. The question was hence how to encourage people to pay to listen to music when it could be downloaded for free. Spotify proved to be the answer. The company was developed by venture capitalists and technologists based in Stockholm. Through their close connections in the Swedish music industry, the small group of entrepreneurs at Rena were among the very first to adopt new technology for the future of the music industry. Their understanding and foresight became the basis for teaching students in a bachelor's programme in music management first offered at Rena in 2005. This programme was unique in Norway and attracted students from across the country. Students who completed their degrees at Rena quickly became attractive to employers due to their knowledge of relevant business models in a renewed music industry.

Swedish partners have also been consistently essential to the developments at Rena. At Falun University College in Dalarna, a high-quality studio was indeed set up at «Lugnet» to produce music. For several years, students from Rena could use this studio as part of their bachelor's program. When this opportunity disappeared, the management at Rena struggled to find other studios because the necessary funding was unavailable at Rena University College. Eventually, agreements were made to send students to Arvika in Värmland as well as to the United Kingdom (UK). The cooperation with Arvika has since deepened, as this locality also seeks to offer a bachelor's programme in music production. Universities in the globally leading music industry nations—the UK and the United States of America—have recently succeeded in persuading Rena to offer streaming and digitalisation courses in English. A course on the role of big data in the music business has also just commenced in the city of Kongsvinger.

Aside from the bachelor's programme, engagement with a wider audience of (especially younger) people interested in music management and production has been critical. Sweden's high level of competence in lifelong learning (including in North Central Sweden more specifically) has been particularly important in this respect (see Table 5.1). Through Interreg's support, actors from Arvika are hired to offer courses in several localities across Hedmark County on the Norwegian side of the border.

More recently, research in music management has been placed on the agenda. Cooperation between Karlstad University and the Inland Norway University of Applied Sciences as well as the campus at Rena has occurred and 10 researchers are now involved on both sides of the border. A centre for music business research operated by Karlstad University and Rena is currently being planned.

The success of music management at Rena would not have been possible without the profound support of Interreg. Management at Interreg in Inner Scandinavia, together with county-level politicians, have looked to the long term when evaluating applications for continued support from Rena every fourth year. Programmes operated by the Norwegian government tend to only look 3–5 years ahead. Management has been clear that commercial music is typically a big city phenomenon, and some of the earlier projects were launched to replicate the advantages of urban agglomeration in a small town such as Rena. A strong belief in the long-term support of creative ideas and entrepreneurs at Rena has existed within Interreg Inner Scandinavia. With the growing importance of Rena and its integration in a global knowledge production system for music, such longevity has added considerable value.

The support of Interreg in building a cross-border innovation system for the music industry, and largely from scratch, constitutes a rare example of path creation in a rural economic fabric. Interreg's support has been knowledge based, including cluster theory and transitioning from a cluster to innovation system. Triple helix cooperation and more recently quadruple helix and ecosystem perspectives have been included. Equal financial risk-sharing between university, college and region has been important.

However, such support is not enough by itself. Rena emerged as a local hub in a global production system for commercial music during a period of creative destruction in the music industry. It can be argued that it is in such periods of rupture and crisis that new windows of locational opportunity appear. Entrepreneurs working as lecturers at Rena College were able to seize opportunities during a period of crisis and combine their resources, first leading to a bachelor's programme in music management. These actors can be termed system-level entrepreneurs who work from Rena and across Sweden and Norway. They quickly established national and global networks with leading actors in Stockholm and Nashville, not only as listeners, but also as partners in dialogue regarding the latest technological and business developments in the industry.

5.5.3 Cleantech Network

The cross-border cleantech network has been developed through Interreg cooperation in Inner Scandinavia since the opening of the fourth generation of Interreg programmes in Europe in 2007. The latest projects, ecoINSIDE1 and 2, are components of the Interreg Norway-Sweden 2014–2020 programme. These projects were developed to contribute to make Inner Scandinavia as a leading cluster within environmentally driven development, and as a cross-border initiative, to strengthen the border region's competitive performance by initiating climate-driven development, reduce border barriers and develop the area's territorial capital.

The application states that this is an ambitious but feasible goal. The key to reaching the goal is the development of the following:

- A world-class border crossing innovation system;
- Inner Scandinavia as an exemplar in sustainable development;
- Binding cooperation between institutions across the border through ownership and greater numbers of shared institutions. (Authors' translation.)

By concentrating on such activities, ecoINSIDE is intended to increase cross-border added value for SMEs, universities and other knowledge providers.

The resources in the project are allocated to three areas: solar energy and energy systems, waste and resources, and sustainable building. The methods used in the project are living laboratories, service innovation, research and development, and innovative public procurement.

Of the 138 SMEs involved in the project, we have interviewed 10. This constitutes a small selection, but it nevertheless provides an overview of the working methods and demonstrates how SMEs benefit from the project and the partnership. Based on the interviews, we have been able to identify the factors that facilitate successful cross-border innovation in the ecoINSIDE project:

- Trust and long-term networks among key partners;
- Experienced project managers as system-level entrepreneurs.

The project managers of ecoINSIDE are professionals with expertise in the three key areas represented in the project: solar energy and energy systems, sustainable building, and waste resources. These project managers operate as system-level entrepreneurs who aim to develop relationships and networks with, among and between producers and users of knowledge by providing connections, knowledge sources and in some cases knowledge itself (e.g. technical know-how and market insights) to organisations in the network.

A group of eight–ten system-level entrepreneurs dedicate between one quarter and one half of their working time to the project. They work with the key partners in ecoINSIDE: Dalarna Science Park (SE), Tretorget (NO), Glava Energy Center (SE) and Kunnskapsbyen Innovation Centre (NO). The latter also acts as the overall project leader. The partners are leading knowledge providers within their areas of competence. Some larger firms, especially utilities, are also partners in the networks.

These organisations and their key employees have learnt to trust each other over a long period of time, having worked together in Interreg and other projects for five to ten years or even longer. In particular, they have learnt to share knowledge and information about two topics: which SME needs what sort of knowledge, and which knowledge providers can offer that particular knowledge? The key partners have several competencies themselves, but their networks include other knowledge providers such as universities and clusters outside of the partnership. Being able to visit SMEs and knowledge providers over a sustained period of time has enabled key personnel to become competent brokers of knowledge between SMEs and knowledge providers in Inner Scandinavia. At times, knowledge is sourced in places such as Stockholm or Trondheim.

These developments would not have been possible without long-term policy support. Interreg officials in Hedmark have worked strategically since the early 2000s to secure funding for a series of projects on clean technology in Inner Scandinavia. Reports from the United Nations (UN) on climate research (International Panel on Climate Change 2007, 2004) have inspired policy-makers. At the same time, they had a strong belief that projects should devise practical solutions. Thus, alongside technical solutions, questions concerning finance, business models and knowledge sourcing have been included in the funded projects. This has helped establish a critical mass of firms and knowledge providers that can generate business and knowledge dynamics within Inner Scandinavia.

5.6 Analysis and Conclusion

The three analysed case studies (Table 5.4) have benefited considerably from the long-term support of Interreg Inner Scandinavia. While the content of the projects has been developed over the years, the same group of key actors has received the majority of the support. This has offered the actors the ability to continue their work to meet rather ambitious goals with content. In all three cases, the ambition has been to operate among the international leaders. The SITE region has the objective of becoming a world-class tourist destination, and in terms of cleantech, Inner Scandinavia is supposed to be an international leader.

The key actors in the three cases also display some interesting differences and similarities. All of the actors share a common interest in starting new activities with the clear goal of seeing results. In other words, they are entrepreneurs. At Rena, actors in the music ecosystem have been affiliated with the college, working either as lecturers or being hired on short-term contracts. In the SITE region, the key actors have been located both within the municipalities and in the private sector. The brokers financed by ecoINSIDE have been able to compensate for some of the disadvantages of working in the Inner Scandinavian region, in terms of establishing an innovation system that works for SMEs and knowledge providers alike (Ørbeck and Braunerhielm 2013).

All three cases are dependent on cross-border cooperation, and such cooperation has come to define the existence and content of the cases. This form of dependence is functional, and the activities on either side of the border are functionally interdependent. They simply could not have existed without each other. The content of the cooperation in the three cases is, however, rather different and the functionality is created by the different types of actors:

Table 5.4 Summarized findings from the three analysed case studies

	Main actors	Role of Interreg	Cross-border functionality	External linkages
SITE tourism development	SITE municipalities and SkiStar	Strategic support since 2005. Finance and knowledge	Learning within winter tourism destination development	Austria
Music ecosystem	HiINN at Rena, Karlstad University, including Arvika College	Strategic support since 2003. Finance and knowledge	Cooperation within music production and management education. Innovations within education	Stockholm, Nashville, UK
Cleantech	Arvika Solar park, Lillestrøm Innovation Centre, Tretorget, Dalarna Science park	Strategic support since 2007. Finance and knowledge	Innovations within solar energy, green constructions and waste handling Includes both knowledge producers and users	Several national and global linkages

- ecoINSIDE: knowledge brokers;
- Music ecosystem: university employees;
- SITE tourism development: managers and politicians within municipalities, and private sector managers.

The existence of SMEs seeking knowledge and universities providing knowledge in a region is in itself insufficient to guaranteeing that a contact to solve a problem will be established. Some sort of arena where two parties can meet and discuss must be made available. Otherwise, the arena has to be constructed, and Interreg demonstrates the potential of constructing such spaces. Flexibility in project design and organisational structure, together with a broad public–private partnership, constitute other important factors that can make positive contributions to the operation of projects. Some of the system entrepreneurs have been based on publicly financed organisations such as municipalities and colleges and can be termed social or political entrepreneurs.

5.6.1 Theoretical Reflections

The proposed ideal types of cross-border integration of regional innovation systems presented in Fig. 5.1 offers a useful point of departure for analysis. The three cases presented above have progressed from being mere weakly integrated systems. In the case of the tourism-based development in the SITE region, it would appear that the major effect is within learning, cooperation and exchange of experiences across the border. Within the music ecosystem and cleantech networks, learning and cooperation are to a greater degree complemented by innovation. The college at Rena is becoming a node in a global knowledge support system within music management. Within cleantech (and especially solar power), Inner Scandinavia has attained a leading position in Scandinavia and also enjoys extensive international channels of knowledge exchange and business development. Based on the ideal types, the three cases are at present semi-integrated systems. The music ecosystem and cleantech are semi-integrated regional innovation systems, while the SITE region constitutes a semi-integrated system of learning and cooperation. Concerning the ideal types, the role of learning and knowledge exchange could have been highlighted to a greater extent as a precondition for cross-border innovation. Contrary to what is often argued in the literature, the three cases demonstrate that new path creation is under certain conditions possible in rural regions. The SITE region was among the pioneers in winter tourism in Scandinavia in the 1960s and 1970s. Both the music ecosystem and cleantech are cases of new path creation, which are not new to just Inner Scandinavia, but in certain aspects also at the national and global scale. The 'secret' behind this new path creation pertains to time and consistent public support.

Indeed, this represents a key finding from the three cases. A history of 15 years of more or less continuous public support through Interreg programmes has been of paramount importance. Interreg has been a mechanism for coordinating innovation policies in the three reported cases. The presence of Interreg projects has been instrumental in helping Inner Scandinavia to evolve into a region with a Stage II, semi-integrated cross-border innovation system. Interreg programmes are based upon the partnership principle and at their best serve as political 'explorer arenas' where new policies can be tested.

Cross-border innovation activities are funded at significantly lower cost for each border region involved due to the co-funding mechanism, hence reducing the financial risk for all parties, including the counties that are involved. Interreg programmes have a 7-year cycle, which is a longer planning horizon than tends to be true of regional or national policy measures. This provides incentives for regional representatives as well as for programme managers to work with complex issues such as path creation, as demonstrated by the case of the cross-border music ecosystem. Aiming for the development of a cross-border, sustainable and world-leading tourism destination (the SITE case) or providing solutions to the existential threat of climate change while promoting technical innovation and seizing new business opportunities are also complex and long-term processes.

A number of the systems entrepreneurs working with consultancies or innovation centres have based their contributions on funding from Interreg projects. Interreg has rendered possible the establishment of numerous arenas for cross-border discussions and cooperation, facilitating learning, innovation and business development.

Functional integration has represented a significant concern in this chapter. We have argued that functionally integrated regions should include much more than the number of commuters. Based on our observations, we note that all three cases have stimulated cross-border functional integration in learning, cooperation and, to some extent, innovation. This was illustrated by a senior officer in one of the SITE municipalities, who claimed to find cooperation within the SITE region more important than with other municipalities in the same county. For knowledge providers and system entrepreneurs within the music ecosystem and cleantech, cross-border functional interdependence has emerged based on numerous years of working together. The projects presented here thus contribute to cross-border functional integration in Inner Scandinavia. More research is, however, required to further specify cross-border innovation and functional integration in rural areas.

Finally, our conclusions add an extra layer of complexity to analysis of regional functional integration processes, and their ultimate contribution to promoting more balanced and cohesive territories, notably in cross-border regions. Indeed, increasing cross-border functional connectivity is a prerequisite for promoting territorial cohesion processes within cross-border regions. This requires, as seen, more that increasing cross-border communing flows. It needs the reinforcement of all sorts of cross-border networking between all involved actors. For this, the EU Interreg programmes, such as Inner Scandinavia, have been playing a fundamental role.

References

Alnes PK, Sæther B (2017) Statistical analysis for the project; Knowledge and innovation strategies involving small and medium-sized enterprises (KISS ME) in the context of INTERREG. Inner Scandinavia. Working paper no. 10, Eastern Norway Research Institute, Lillehammer

Asheim BT, Gertler MS (2005) The geography of innovation: regional innovation systems. In: Fagerberg J, Mowery DC, Nelson RR (eds) The Oxford handbook of innovation. Oxford University Press, Oxford

Boschma R (2005) Proximity and innovation. A critical assessment. Reg Stud 39:61–74

Danermark B, Ekström M, Jakobsen L, Karlsson JC (2002) Explaining society. Critical realism in the social sciences. Routledge, London

European Commission (2018) Proposal for a regulation of the European parliament and of the council on the specific provisions for the European territorial cooperation goal (Interreg) supported by the European regional development fund and external financing instruments. COM 374, final, Strasbourg

Gundersen F, Jukvam D (2013) Inndeling i senterstruktur, sentralitet og BA-regioner. Report no. 1, Norsk Institutt for by- og regionforskning, Oslo

Isaksen A, Karlsen J (2016) Innovation in peripheral region. In: Shearmur R, Carrincazeaux C, Doloreux D (eds) Handbook in the Geographies of Innovation. Edward Elgar, Cheltenham, pp 277–285

Jakobsen SE, Lorentzen T (2015) Between bonding and bridging. Regional differences in innovative collaboration in Norway. Nor J Geogr 69:80–89

Karlsen J, Larrea M (2014) Territorial development and action research. Innovation through dialogue. Routledge, London

Lundquist KJ, Trippl M (2013) Distance, proximity and types of cross-border innovation systems: a conceptual analysis. Reg Stud 47:450–460

Maggioni MA, Uberti TE (2007) Inter-regional knowledge flows in Europe: an econometric analysis. In: Frenken K (ed) Applied evolutionary economics and economic geography. Edward Elgar, Cheltenham, pp 230–255

Medeiros E (2014a) Is there a new TRUST in Inner Scandinavia, evidence from cross-border planning and governance. Geogr Ann: Ser B, Hum Geogr 96(4):363–386

Medeiros E (2014b) Territorial cohesion trends in Inner Scandinavia: the role of cross-border cooperation (INTERREG-A 1994–2010). Nor Geogr Tidsskr 68(5):310–317

Medeiros E (2017) Cross-border cooperation in inner Scandinavia: a territorial impact assessment. Environ Impact Assess Rev 62:147–157

Nauwelaers C, Maguire K, Marsan GA (2013) The case of Hedmark-Dalarna (Norway-Sweden)—regions and innovation: collaborating across borders. OECD Regional Development Working Papers no. 18, OECD Publishing Paris

Ørbeck M, Braunerhielm L (2013) Grenseindeks og grenseanalyser Interreg Sverige – Norge. Rapport no. 7. Østlandsforskning, Lillehammer

Ragin CC, Amoroso LM (2011) Constructing social research, 2nd edn. Sage, Los Angeles

Rypestøl JO (2018) New regional industrial path development. Entrepreneurs, knowledge exchange, and regional contexts. Ph.D. thesis, University of Agder

Tomaney J (2009) Region. International encyclopedia of human geography, pp 136–150

Tödtling F, Trippl M (2005) One size fits all? Towards a differentiated regional innovation policy approach. Res Policy 34:1203–1219

Bjørnar Sæther is professor in Human Geography at the Department of Sociology and Human Geography at the University of Oslo. His research and teaching activities are within economic geography and includes innovation in rural areas, rural development, natural resource based industries and related questions on sustainability. He teaches on economic globalisation and the geography of innovation and sustainability transitions.

Bjørn Terje Andersen has been programme manager from 1997 to 2016, from the early beginning of Interreg Inner Scandinavia 1997–1999 and later on Interreg Sweden-Norway programme. Development of structures among secretariat, steering committees and programme management together with Country Administrative Board in Värmland. Development of 7 years of cross-border programme 2000–2007, 2007–2013 and this period 2014–2020 in cooperation with regional partnership along the border. In the last period building up methodology in a national web module how to apply funding from Interreg Sweden-Norway, methodology how to make an assessment of the Project Applications in Interreg Sweden-Norway for the period 2014–2020. He also worked in evaluation, project management. He holds a bachelor's degree in business management and audit control from University of Agder, and management programme from BI Norwegian Business School.

Erik Hagen is a Program Manager at the Interreg Sweden-Norway 2014–2020 programme's Joint Technical Secretariat. He is a graduate from the Norwegian University of Life Sciences. He is experienced in public administration from the local to the national level and has held leading positions in the private and voluntary sectors. In Interreg Cross Border Cooperation, he has been working on cross-border data, research, analysis and strategies, coordinating the development of the Interreg IV (2007–2013) and the Interreg V (2014–2020) generations of the Interreg Sweden-Norway programme. Since 2003, he has co-developed a vast number of cross-border initiatives in a variety of policy areas including innovation, energy and climate change, cultural exchange, education and nature conservation. As a Program Manager, he is responsible for Interreg funds being invested in an efficient and transparent manner. In co-work with the academia, he has piloted Territorial Impact Assessment for Cross Border Cooperation and established a series of other basic research studies regarding borders. He has followed closely the European policy development processes for cross-border regions via the Association of Cross Border Regions (AEBR) and the Mission Opérationnelle Transfrontalière (MOT). From 2018, he is the ESPON Contact Point (ECP) for Norway.

Chapter 6
Integrated Sustainable Urban Development Strategies in the European Union: Added Value and Challenges

Arno van der Zwet and Martin Ferry

Abstract This chapter will consider the implementation of integrated sustainable urban development strategies as part of this chapter requirements under the 2014–2020 European Regional and Development Fund regulation. This chapter will reflect on the place-based rationale of the approach and consider the major innovations in the 2014–2020 regulations. It will subsequently consider how Member States have designed and implemented the regulation, particularly focusing on variation between and within member states. The second section of the chapter considers the added value of the provisions at the European level. Added value can be captured in three dimensions: the extent to which new or strengthened strategic frameworks have emerged, the extent to which integrated governance and strengthened implementation capacities have been achieved and the extent to which experimentation and innovation in relation to interventions have taken place. The third part of the chapter will analyse some of the key challenges in relation to implementation of these strategies through European funding streams. These relate to issues around capacity, regulations and governance. The final section will reflect on the lessons that can be learned in relation to the role of integrated place-based strategies to achieve territorial cohesion.

Keyword Integrated sustainable urban development · European Union
EU Cohesion Policy · Urban development · Territorial cohesion

---

A. van der Zwet (✉)
University of the West of Scotland, Paisley, Scotland
e-mail: Arno.van-der-Zwet@uws.ac.uk

M. Ferry
European Policies Research Centre, University of Strathclyde, Glasgow, Scotland
e-mail: martin.ferry@strath.ac.uk

E. Medeiros (ed.), *Territorial Cohesion*, The Urban Book Series,
https://doi.org/10.1007/978-3-030-03386-6_6

6.1 Introduction

The following chapter considers the implementation of integrated sustainable urban development strategies at the European level. These strategies have been introduced in the 2014–20 Cohesion Policy as part of a shift to place-based policy thinking and practice (van der Zwet and Bachtler 2018). In particular, the urban dimension of Cohesion Policy was strengthened in the 2013 reforms of the European Structural and Investment Funds (ESIF) for the 2014–20 period (Tosics 2015). Developments in thinking about place-based approaches (van der Zwet and Mendez 2015) were particularly influential in the debate on reforming Cohesion Policy in the mid/late 2000s and were given credence by a number of reports (Barca 2009; Farole et al. 2009; OECD 2009a, b). The 2009 Barca Report argued that such policy interventions are superior to spatially blind interventions, which too often assume a top-down approach (Barca 2009). In essence, integrated place-based approaches rely on local knowledge, capital and control over resources as well as a locally developed strategic framework in order to facilitate endogenous growth.

Territorial provisions have played a relatively small but significant role in previous programme periods of EU Cohesion Policy. For example, the Urban Community Initiative, first launched in the 1994–99 period, continued in the 2000–06 period and integrated in the Investment for Growth and Jobs programmes in the 2007–13 period, encouraged urban areas and neighbourhoods to design innovative, integrated urban development measures. Under the European Territorial Cooperation programme, URBACT was set up in 2003 and has sought to foster sustainable integrated urban development in cities across Europe. URBACT is mainly a knowledge exchange platform, enabling networking between cities and identifying good practice. The LEADER approach was established in 1991 and has become an important element of rural development, and since 2007 it has also been used within the European Maritime and Fisheries Fund (EMFF) to support sustainable development in fishing communities. Community-Led Local Development (CLLD) was introduced for the 2014–20 period, based on the LEADER instrument.

The new emphasis on integrated place-based approaches under Cohesion Policy in the 2014–20 period follows the formalisation of territorial cohesion as an objective for the EU in the Treaty on the Functioning of the EU (TFEU) and the subsequent regulations for European Structural and Investment Funds approved in 2013. According to the Territorial Agenda of the European Union (European Commission 2011), territorial development policies should address the following issues:

- Increased exposure to globalisation and structural changes caused by the global economic crisis;
- New challenges for European integration and growing interdependence of regions, territorially diverse demographic and social challenges and spatial segregation of vulnerable groups;
- Climate change and environmental risks that have geographically diverse impacts;
- Growing energy challenges threatening regional competitiveness; and

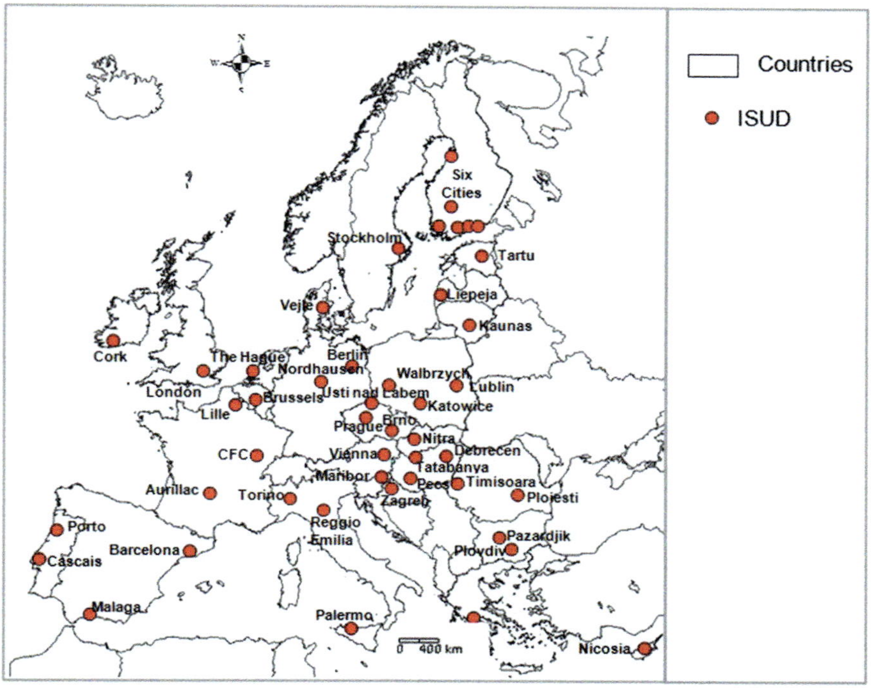

Fig. 6.1 Location of in-depth ISUD case studies. *Source* own elaboration

- Loss of biodiversity and growing vulnerability of natural, landscape and cultural heritage.

The nature of these challenges is thought to require an integrated mix of interventions in order to increase their impact and to exploit fully the development potential of different types of territories. There is a particular focus on fostering Integrated Sustainable Urban Development (ISUD) through integrated strategies in order to strengthen the resilience of cities.

This chapter draws on research that was undertaken as part of a European Commission (EC) study of the integrated territorial and urban strategies supported by European Structural and Investment Funds (van der Zwet et al. 2017). The research is the most extensive data gathering exercise in relation to ISUD strategies that have been carried out to date. The project identified 853 ISUD strategies and a further 153 territorial strategies that are implemented by integrated territorial investment tool but that are not considered part of each Member State's ISUD allocation. A second stage involved the creation of a database of 426 strategies and the development of 42 in-depth ISUD case studies (Fig. 6.1). These case studies involved in-depth semi-structured interviews with Stakeholders in 26 Member States.[1] This

[1] Malta and Luxembourg were not included.

chapter will only discuss ISUD strategies and mainly draw on the evidence collected as part of the case study fiches. Table 6.1 provides an overview of the in-depth case studies. The table reports the urban area, Member State whether the strategy covers a metropolitan area, town or neighbourhood. The total population coverage of the strategy area, the implementation mechanism (see next section)—either Integrated Territorial Investment (ITI), Priority Axis (PrAxis) or Operational Programme (OP) and whether the urban area is located in a More Developed region (MD), Transition Region (TR) or Less Developed region (LD).

The remainder of this chapter is structured as follows. The next section will provide an overview of the key aspects of integrated sustainable urban development in the 2014 ESIF period. In the next section, we will consider the potential added value of ISUD and other territorial approaches. In the penultimate section, we shall consider some of the challenges of implementing ISUD and other strategies. These include capacity, regulatory and governance issues.

6.2 Overview of Integrated Sustainable Urban Development Strategies in the 2014–2020 ESIF Period

In the 2014–20 programme period there are a number of important differences compared to previous periods. First, the overall funding allocation for integrated place-based approaches has increased. According to the indicated territorial delivery mechanisms in the OPs, around 9% of the Cohesion Policy budget (EUR 31 billion) will be spent through the various territorial provisions. Second, there is a regulatory requirement to implement integrated place-based approaches in cities. Third, the integrated approach, in general, is emphasised. Fourth, more information regarding the implementation of integrated place-based approaches is required at the programme level. Fifth, there is more attention for knowledge diffusion (e.g. providing guidance, scenarios, participation in urban networks, peer-to-peer review, etc.).

The use of this chapter for the implementation of European Regional and Development Fund (ERDF) makes integrated urban development a compulsory feature of the ESIF regulation. One of the main goals of the approach is to empower cities. As such, a novel feature of the regulation is the requirement to delegate implementation tasks to cities for interventions that are programmed as part of the minimum 5% ERDF share to implement ISUD. However, it is left to the Member States to identify those territories that are considered urban areas. Furthermore, the regulation encourages the introduction of innovation and experimentation (Urban Innovative Actions, Chap. 7 of Regulation 1301/2013) and the introduction of an Urban Development Network to deepen the discussion on the implementation of the urban dimension (Chap. 8 of Regulation 1301/2013).

This chapter can be implemented using a number of different approaches and instruments. ISUD can be implemented through so-called mainstream approaches (i.e. in a similar way to how other ESI Funds are implemented) as either a separate

Table 6.1 In-depth ISUD case studies

Case study	Member state	Type of city/region	Population	Implementation method[a]	Type of region[b]
Vienna	AT	Metropolitan	1,840,000	PrAxis	MD
Brussels	BE	Town	1,139,000	OP	MD
Plovdiv	BG	Town	504,338	PrAxis	LD
Pazardjik	BG	Town	69,384	PrAxis	LD
Nicosia	CY	Neighbourhood	8244	PrAxis	TR
Prague	CZ	Metropolitan	609,000	ITI	MD
Brno	CZ	Town	2,000,000	ITI	LD
Ústí nad Labem	CZ	Town	52,000	ITI	LD
Berlin	DE	Metropolitan	3,500,000	PrAxis	MD
Nordhausen	DE	Town	41,839	PrAxis	TR
Vejle	DK	Town	53,230	PrAxis	MD
Tartu	EE	Metropolitan	120,929	PrAxis	LD
Patras	EL	Neighbourhood	150,000	ITI	LD
Malaga	ES	Town	59,695	PrAxis	TR
Barcelona	ES	Town	114,014	PrAxis	MD
Six cities	FI	Town	1,600,000	ITI	MD
Aurillac	FR	Other	54,036	PrAxis	TR
Centre Franche-Comté	FR	Region	319,868	PrAxis	TR
Lille	FR	Metropolitan	357,220	ITI	MD
Zagreb	HR	Town	1,086,528	ITI	LD
Pecs	HU	Town	145,000	PrAxis	LD
Debrecen	HU	Town	145,000	PrAxis	LD
Tatabanya	HU	Town	68,000	PrAxis	LD
Cork	IE	Metropolitan	119,230	PrAxis	MD
Torino	IT	Town	905,000	OP and PrAxis	MD
Palermo	IT	Town	1,069,754	OP and PrAxis	LD
Reggio Emilia	IT	Region	171,655	PrAxis	LD

(continued)

Table 6.1 (continued)

Case study	Member state	Type of city/region	Population	Implementation method[a]	Type of region[b]
Kaunas	LT	Neighbourhood	297,846	ITI	LD
Liepaja	LV	Town	71,926	ITI	LD
The Hague	NL	Town	510,000	ITI	MD
Katowice	PL	Metropolitan	2,759,961	ITI	LD
Walbrzych	PL	Metropolitan	415,800	ITI	LD
Lublin	PL	Metropolitan	547,784	ITI	LD
Porto	PT	Metropolitan	237,534	ITI	LD
Cascais	PT	Town	206,479	PrAxis	MD
Timisoara	RO	Town	387,000	PrAxis	LD
Ploiesti	RO	Town	327,000	PrAxis	LD
Stockholm	SE	Metropolitan	2,100,000	OP	MD
Maribor	SI	Town	81,165	ITI	LD
Nitra	SK	Town	92,935	ITI	LD
London	UK	Metropolitan	8,539,000	ITI	MD

Source van der Zwet et al. 2017
[a]Integrated Territorial Investment (ITI), Priority Axis (PrAxis), Operational Programme (OP)
[b]More Developed region (MD), Transition Region (TR), Less Developed region (LD)

OP or a separate mixed priority axis. ISUD can also be implemented through an Integrated Territorial Investment (ITI) strategy. This new tool provides a framework for thematic/sectoral integration and can be used to combine resources from different funds (usually ERDF and European Social Fund (ESF)) into single strategies. ITI can also be used for territorial strategies that do not contribute to this chapter regulatory requirements. These type of strategies are referred to as regional ITI and although they are in many ways similar to ITI ISUD strategies (i.e. they are integrated, drawing funding from multiple priority axis and/or funds and have a dedicated ring-fenced budget), they are particularly different in terms of the afforded responsibilities in relation to project selection to the local level which is in all cases more limited.

Community-Led Local Development (CLLD) can also contribute to ISUD strategies. CLLD provides a bottom-up participatory approach to ESIF implementation generally and can also be used in the urban context. However, ITI and CLLD have a broader application. ITI can also target functional areas, such as rural, rural–urban and cross-border areas and territories with specific geographic features (van der Zwet et al. 2014). CLLD can also contribute to the implementation of these non-ISUD ITI strategies.

These measures all aim to support place-based development in an integrated manner across the EU28. However, overall, the regulatory framework, particularly in relation to this chapter, is not prescriptive and provides extensive scope for variation in implementation across the EU28. This flexibility is considered a strength as it allows for more place-specific interventions. However, the Commission guidance lists some of the following key principles for territorial provisions (EC 2014):

- Include a comprehensive and evolving strategy that is of real use to the urban authority;
- Include a robust territorial and demographic analysis;
- Include a mid-term/long-term vision, i.e. until at least 2020;
- Include a system of interlinked actions which seek to bring about a lasting improvement in the economic, environmental, climate, social and demographic conditions of an urban area;
- Build upon other major investments in the urban area;
- Be coherent with the overall development targets of the region and Member State;
- Be realistic in terms of the capacity to implement;
- Be linked to the objectives of the programme from which the funds derive; and
- Demonstrate how local citizens, civil society and other governance levels will be involved in the implementation of the strategy.

These core principles already highlight where the added value of the approach can be expected and also where we can anticipate some of the challenges in terms of design, implementation and governance of the strategy. These issues will be further discussed in the next two sections.

6.3 The Added Value of Territorial Approaches

Analyses of the influence of Cohesion Policy in changing the policy and practices of regional and urban development in Member States are often discussed under the broad heading of 'added value' (Bachtler et al. 2009). These analyses have highlighted changes in the way that practitioners and stakeholders conceptualise and relate to regional policy through involvement in Cohesion Policy programmes: in the content of the policy (strategic goals, underpinning rationales and measures), and in the way policy is designed and delivered. Generally, Cohesion Policy is credited with adding value in a number of ways. First, it can support and strengthen the profile and strategic framework of regional policy in Member States (Mairate 2006). This can involve a raised awareness among key actors of the role of strategy building. It can also build capacities that are durable beyond the project level and facilitate future project implementation at different scales. Cohesion Policy can encourage information and knowledge exchange on key strategic priorities, creating alignments and synergies between different levels of government and across administrative boundaries. Taken together, Cohesion Policy concerns the creation of new strategic frameworks and/or the strengthening of pre-existing approaches across different territorial levels in Member States.

Second, Cohesion Policy can encourage integrated governance and strengthen capacities within member states (EC 2016). This process consists of a number of aspects as given below:

- The establishment of new structures, arenas and partnerships for strategic thinking in the territory;
- Building up social capital 'soft' skills, consensus building and trust building;
- Developing technical skills and capacity at local level; and
- Providing input into policy development and policy instruments.

Third, Cohesion Policy can promote experimentation and innovation, with interventions facilitating greater cooperation and collaboration among policymakers and stakeholders at different levels (Bache 2010). As such, the policy increases awareness of opportunities that aid development in the territory. It can create investment-steering and investment-accelerating effects. It can leverage financial and 'other' incentives to mobilise actions and resources. This can produce multiplier effects, by encouraging the 'pooling' of regional policy budgets and administrative resources, accessing additional funding for regional development from public and private sources and increasing the effectiveness and impact of regional policy by strengthening coordination and synergies between EU and domestic instruments.

In this way, ESI Funds act as motivators or 'agents of change' (Polverari et al. 2017). Although at an early stage of implementation, the introduction of ISUD strategies creates substantial potential[2] for the creation of these dimensions of EU 'added value'. The next subsections will identify examples of each of these added value dimensions at the urban level, providing examples where appropriate.

6.3.1 New or Strengthened Strategic Frameworks

The process of developing ISUD strategies have already created added value by demonstrating to stakeholders in the territory the role and significance of integrated strategic approaches. Potential added value is recognised in addressing inefficiencies caused by fragmentation. The involvement of local authorities in the design and implementation of strategies is credited by implementing authorities with creating potential for minimising rivalry, competition and duplication of projects. In Lublin, for instance, the development of the ISUD strategy has increased the knowledge and awareness of the role and importance of strategic and integrated programming. The standard of strategic planning for development has increased and local authorities have become much more involved in Cohesion Policy implementation (as opposed to acting only as beneficiaries).

Thus, there is a clear process of local-level capacity building underway. In the Patras ISUD strategy, a key component of perceived added value among policymakers is that, contrary to previous programme periods, which relied on project-based and fragmented interventions, the strategy now sets out an integrated plan with a particular geographical focus. Similarly, in the Prague ISUD strategy, there is a

---

[2]It is important to stress that, at this stage, the added value is often only *potential*; new frameworks and mechanisms have been introduced, but their operation in practice is not tested or assessed.

presumption that the adoption and implementation of the strategy will ensure better functional links between constituent areas, by developing strategic solutions to common problems. In the Kaunas ISUD strategy, local authorities see the strategy planning and implementation process as a good exercise to prove the use of integrated planning in the real life. Success of the strategy will be an important determinant of whether and to what extent an integrated approach will be introduced into city planning in the future.

Some areas have long-established traditions of working with integrated place-based strategies and limited ESI funding allocations. In these cases, added value can be identified in the extension or strengthening of existing practice. For example, the Cork ISUD strategy strengthens the integration of the country's overall approach to regional development with local development plans. It offers the opportunity to fund a range of projects and embed them into an integrated plan for the city. From the city-level perspective, the link between the city plan and ESI funding provides the opportunity to 'think bigger' and more strategically about which projects they want to fund.

6.3.2 Integrated Governance, Strengthened Implementation Capacities

The implementation of ISUD strategies is creating added value in some contexts in the form of new, cooperative governance mechanisms and structures. The establishment of intermediary bodies as part of this chapter requirements is, in some cases, having an observable influence on how interventions/projects are implemented. This chapter stipulates that the input of urban authorities in resource allocation decisions must be demonstrated, particularly in the selection and delivery of projects. This secures active participation in resource allocation decisions, and in many cases is accompanied by new systems, structures and tools that maximise the input from partners and stakeholders. In some cases, implementation by urban authorities has required organisational arrangements that increase resources for implementation, potentially boosting capacity in the longer term.

These new approaches to governance can include the development of different governance structures, processes and capacities that cover different types of functional areas. The Brno ISUD ITI strategy, for instance, has become a catalyst for institutional changes in metropolitan cooperation and has enabled wide agreement on, and funding for, strategic projects principally for the metropolitan territory. There are now efforts to ensure the continuation of the structures created (e.g. steering committees, working groups) and metropolitan partnerships.

In the Lublin ISUD strategy, the added value of ITI is seen as substantial by the MA, the ITI Office and the ITI partners in changing approaches to territorial governance in the region. The ITI strategy has created a governance framework that incentivises an integrated approach to territorial governance. City and local authorities are working together on the ITI strategy and are trying to use this cooperation for the development of the whole area. It is worth noting that before signing the agreement in 2014, neither the mayors nor the operational civil servants of the Lublin municipalities were in regular contact with each other to discuss strategic development. Thanks to a special model of ITI cooperation which includes an operational ITI strategy coordinator in every partnering municipality, officials are in contact on a daily basis, while the mayors meet at least once a month to discuss more strategic issues. Such close interaction would not have happened without the ITI strategy framework and the associated incentives.

Strengthened cooperation can also concern partners from different sectors. Integrated place-based strategies involve a much broader range of actors compared to simple projects, and this can strengthen social networks based on reciprocity, trust and cooperation. For example, in the Maribor ISUD strategy, added value is identified in the intensive cooperation with the university and NGOs. Furthermore, a feature of added value noted in the case of the Brussels ISUD strategy is the development of partnerships, including those involved in the social economy and voluntary sectors.

6.3.3 Experimentation, Innovation with Interventions

Specific features of ISUD strategies also increase the potential for experimentation with new and innovative approaches to designing and delivering initiatives, in turn creating added value. For instance, in some strategies the scope to combine different ESI funds is seen as a source of added value, providing efficiency gains from exploiting synergies between different funding streams in one integrated place-based strategy.

Implementing authorities for the 'The Hague' ISUD strategy highlight the value of integrating ESF and ERDF funding in the territory. It is too early to fully understand how effective this approach is, but it does encourage policymakers and project stakeholders to at least think in a more integrated way. The integration at the level of the ITI strategy is considered a first step towards further integration at the project level. It is noted that there are important differences in terms of culture, implementation practices and types of stakeholders between the funds, which form a significant barrier to full integration. However, by combining ERDF and ESF within an ITI strategy, these barriers are bridged by both public administration bodies and stakeholders.

In other cases, policymakers are taking advantage of ESIF ISUD strategies to pilot new configurations of territories and stakeholders, including private sector partners.

The Finnish Six Cities ISUD strategy represents an innovative type of operational cooperation between the six cities, which has emerged from their needs (i.e. joint interests and measures). The starting point was that the strategy would not just entail one or two cities, but multiple cities across Finland. The instrument is perceived to be valuable and innovative as it also promotes cooperation with businesses and strives to achieve other objectives such as competitiveness and growth. Added value is also achieved by increasing awareness of investment opportunities and the formation of links with the private sector that can facilitate private funding for specific, innovative types of actions.

In the Vejle ISUD strategy, the expected added value is that it will help to build a common basis for public–private partnership and in so doing strengthen cooperation on sustainable urban development. For example, from a small project on the utilisation of construction waste, it is expected that awareness will be strengthened among SMEs of the business potential in the more sustainable utilisation of waste.

6.4 Design and Implementation Challenges

Most research, studies and evaluations of Cohesion Policy implementation and management identify a number of challenges (e.g. Dotti 2016; Bachtler et al. 2014; McCann and Ortega-Argilés 2016). Although the challenges are generally very diverse in nature, they can be captured in a number of interlinked dimensions as follows:

- Resource, institutional and administrative capacity;
- Regulatory challenges; and
- Governance challenges.

Each of these dimensions is captured in Fig. 6.1 under which we can subsequently identify sub-dimensions and specific challenges.

6.4.1 Resource, Institutional and Administrative Capacity

Earlier studies have raised concerns about resource, institutional and administrative capacity to manage and implement strategies, particularly where responsibilities for implementation are delegated to local bodies with more limited expertise or resources to implement ESIF funds (Bachtler et al. 2014). These concerns about capacity are also linked to the perceived increase in the complexity of the ESIF regulations, sometimes due to 'gold-plating' by Member States rather than the original regulations (van der Zwet et al. 2014; Böhme et al. 2017).

In first instance, city authorities from, for example, Berlin, Chomutov and Lille, have noted that the design phase of strategies can be very lengthy which can drain resources and is often perceived as a cumbersome and bureaucratic process and in some cases hindered by lack of or complicated guidance. Institutional capacity is also negatively affected either by tight deadlines (e.g. Cascais, Porto) or by processes that are considered too lengthy (e.g. Barcelona—See Fig. 6.2), and by strategies that have suffered from delays and overlapping processes (e.g. Debrecen, Ploiesti, Tatabanya and Timisoara) that influence the quality of design and the speed of implementation.

These capacity challenges can include those at the institutional level, which reflect issues around the administrative burden for local authorities. Challenges may relate

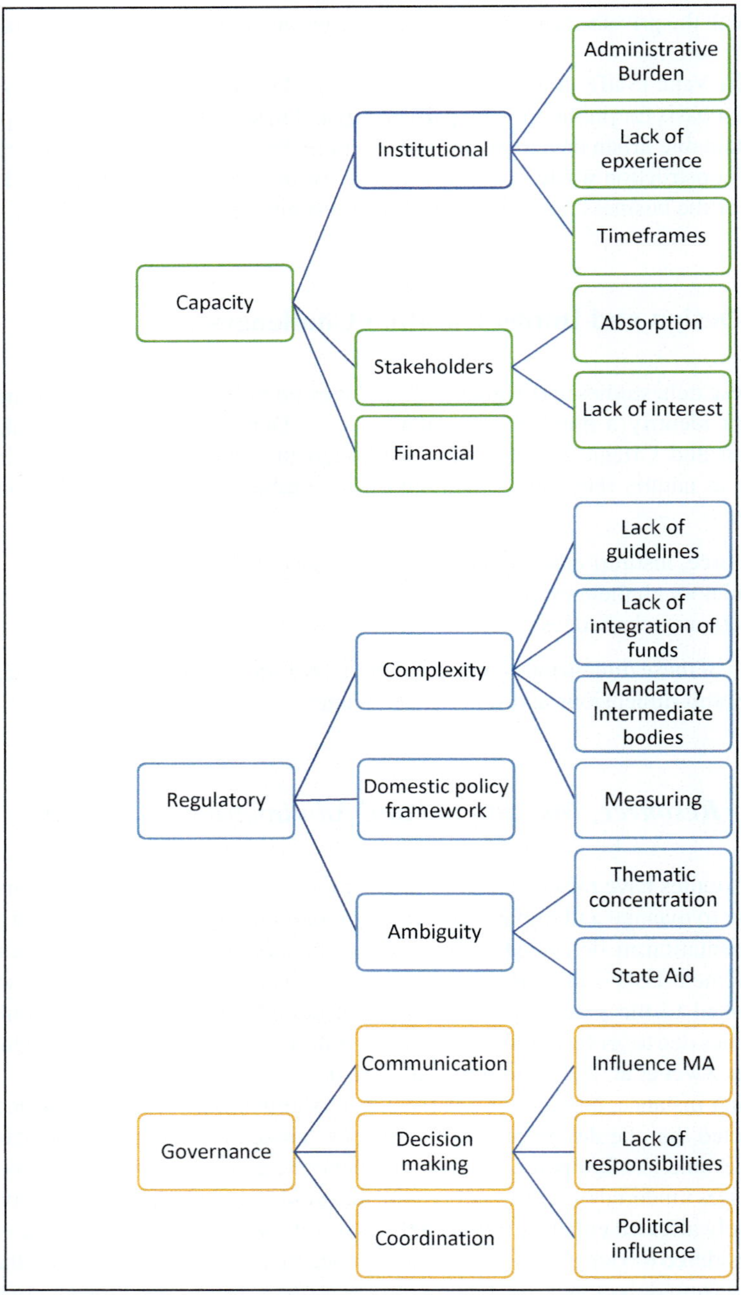

Fig. 6.2 Challenges to implementing ISUD. *Source* van der Zwet et al. (2017)

to unfamiliarity with the implementation of ESIF funds, but also limited capacity in implementing place-based integrated strategies. This refers in particular local administrations that are small and have limited experience in implementing these types of initiatives. From a local perspective, issues of capacity are often linked to what is considered excessive complexity, particularly in relation to the governance of the strategies but also the regulatory or other formal demands (i.e. country-specific guidance) placed on local actors. Capacity challenges also occur at the Managing Authority (MA) level, where the introduction of integrated place-based approaches has added to an already heavy administrative burden.

Second, beneficiaries and stakeholders may lack capacity in terms of experience with ESIF projects. The design and implementation of integrated place-based approaches are in many cases significantly different from previous approaches and includes different beneficiaries and stakeholders. This lack of capacity and understanding can lead to disinterest. The inclusion of certain stakeholders in the design process proved challenging in some strategies. Despite extensive efforts, public engagement in the design process was considered to have limited success in Debrecen and Pecs. In Kaunas, the short timescales afforded limited opportunities for engagement with stakeholders. Beneficiary recruitment can also be challenging, either because new groups are targeted (e.g. Brussels, Six Cities) or because of a scarcity of suitable beneficiaries that can absorb funding (e.g. Debrecen).

Third, capacity challenges are also linked to the ability to implement the strategy due to limited funding. There are also concerns in relation to the scale of funding allocations, dispersion of responsibilities and funding. In most Member States, the level of funding allocated to ITI (and integrated place-based approaches more broadly) is relatively limited, raising questions about their potential impact (van der Zwet et al. 2014). On the one hand, budgetary restrictions can limit the scope of a strategy or conversely make the implementation of a comprehensive strategy unrealistic. Discrepancies between the aims of strategies and the funding that is required to achieve are not necessarily problematic, as it can lead to effective prioritisation and better understanding of the strategic choices that need to be made by a wider group of stakeholders. However, it can lead to tensions between stakeholders within the territory as well as between different levels of government. Furthermore, in cases where financial allocations are small, either because the overall Member State allocation is small or because funding has been dispersed over many territories, the effectiveness and efficiency of the approach can be questioned.

An overreliance on ESIF funding is in some cases also considered problematic. Additionally, the distribution of reduced funds over a broader array of priorities can cause fragmentation. Domestic budgetary restrictions can cause challenges in terms of securing co-financing, which in some cases can impact on design. Limited funding may also influence the design of indicators, as the funded operations are unlikely to have a major impact that can be measured using common indicators.

6.4.2 Regulatory Challenges

A second overarching category can be described as regulatory challenges. These relate in the first instance to a perceived complexity. Some of the evidence suggests that there is an inherent tension between, on the one hand, the flexibility afforded to Member States in terms of the different ways in which integrated place-based approaches can be implemented under the 2014–20 framework particularly in relation to the different mechanisms that can be used, the diverse range of territories that can be targeted and the integration with domestic implementation structures. This flexibility is considered valuable and positive and allows Member States, regions and urban authorities to adopt approaches that are sensitive to the context. On the other hand, this flexibility means there is a certain amount of ambiguity in relation to the rules and regulations (van der Zwet et al. 2014).

However, it can also be associated with a lack of capacity and lack of understanding of integrated place-based approaches, which can lead to calls for more guidance. In these cases, the lack of—or late provision of—guidelines is closely linked to the perception of complexity (see previous section). In this context, it is important to distinguish between EU and domestic guidelines in this case and recognise the knock on effects that may occur. In most cases, it is the domestic guidance that is considered more problematic, but the delays in domestic guidance were often a consequence of the late approval of guidance at the EU level.

Late provision of guidance can be particularly challenging in those cases where strategy design had already started and had subsequently be adapted (e.g. Brno, Chomutov, Patras, CFC pole) or could not inform the full design process (e.g. Tatabanya, Maribor). In some cases, a continued absence of guidance at the domestic level is considered to have had a negative impact on the quality of the strategy. Guidance can also be considered too restrictive and leading to an approach that is too uniform (e.g. Kaunas) or too complex (e.g. Vejle).

Another element that is in some cases linked to the complexity issues relates to the measurement and development of meaningful indicators. Several urban authorities and MA report a lack of data sources on which a comprehensive area analysis could be based, particularly at neighbourhood level, but also in some cases at city level (e.g. Kaunas, Lublin, Ploiesti and Zagreb). More fundamentally, a high number of urban authorities and MA consider the existing Cohesion Policy indicator framework to be inappropriate in relation to ISUD. The vast majority of indicators are sectoral and fail to capture the integrated territorial impact of strategies (e.g. CFC pole, Kaunas, Maribor, Nitra, Pazardzhik, Pecs, Plovdiv and Prague).[3]

A further specific issue in relation to complexity is that the 2014–2020 regulatory framework provides scope for the integration of ESIF funds at the strategic level but opportunities for meaningful integration of funding streams at the project level remain very limited (e.g. Katowice, Brno, Six Cities, Kaunas, Stockholm and The Hague). This point is particularly made in the Katowice case, where it is noted that there is a need for more clarity and flexibility in the rules and guidelines for

[3] For more information see Ferry et al. 2017.

implementation produced by the Commission, including on how to plan integrated projects.

According to several urban and managing authorities, the mandatory requirement for urban authorities to be designated as intermediate bodies has created unnecessary complexity. For example, in Brno the diverse implementation structure for the ERDF flows (due to a mandatory requirement for an Intermediary Body (IB)) on the one hand, and for the ESF and Cohesion Fund (CF) on the other hand, complicates the implementation mechanisms of ITI. In some cases, it has been urban authorities who have been concerned with being designated formal IB status as it increased their workload but also some managing authorities have been concerned with specific capacity challenges at the local level. Conversely, some managing authorities and urban authorities have recognised that despite initial difficulties the introduction of the intermediary body requirement has led to capacity building at the urban level (Ferry et al. 2018).

The lack of a domestic urban policy framework or sufficient linkages to domestic policy frameworks can also hamper effective design and implementation. For example, a key problem for Turin is represented by the lack of a national urban strategy and by the fragmentation of responsibilities for urban development at the national level, which means that cities must interact with different ministries/agencies.

There are also inconsistencies and ambiguities within the ESIF framework with regard to supporting integrated place-based approaches, particularly the requirement for thematic concentration is on occasions at odds with an integrated approach. In some cases, urban actors note that the decisions on the themes that are covered by the ESIF programmes and which are informed by the thematic concentration principle mean that not all themes that relate to the local needs of strategies are covered in the programmes, and therefore cannot be covered in the strategy. This requirement either meant that urban authorities responsible for the development of the strategies were forced to adopt themes that were not considered a priority or they could not include themes that were a priority.

For example, in Brno, the gradual narrowing of eligible themes and activities from national level for the ITI strategies has undermined the confidence of local partners in the capabilities of the ITI instrument. Also, in Chomutov, it is noted that only part of the strategy's scope can be implemented due to thematic narrowing. Tatabanya also deemed that a greater diversity of interventions was necessary. Conversely, in the Lille strategy, the urban authority wanted a narrower focus, whereas the MA wanted to cover all four axes of the programme. In Pecs, the strategy formulation started on the assumption that it would encompass territorial and sectoral measures. However, the menu system and the predefined breakdown of funds as well as eligibility provisions altogether inhibited the use of a truly integrated approach at both project and programme levels. General issues of aligning strategies to programme priorities have also been noted in the Six Cities strategy in Finland where there have been some challenges to ensure that the cities 'understand' how to align the implementation of the strategy so that it contributes to the overall objectives of the OP.

A final set of regulatory challenges relate to ambiguities in relation to the wider EU regulatory framework, in particular, concerning state aid requirements, which limit the implementation of strategies. For example, in Aurillac urban regeneration projects focusing on housing, the revival of retail activities or sustainable mobility, usually require a public–private joint venture because of their size and complexity, especially in a context of limited public finances. However, they face state aid restrictions. In this context, many urban and managing authorities note the inconsistencies with regards to state aid rules at the commission level as creating legal uncertainty.

6.4.3 Governance Challenges

A third category of challenges falls under the broad heading of governance. Integrated place-based approaches require intensive coordination between different levels and different policy areas, which presents its own challenges in terms of planning. Barca also warns of potential failure of coordination, leading to an underprovision of some public goods and services and overprovision of others (Barca 2009). The experiences from several ISUD strategies demonstrate that these challenges can relate to issues of communication, particularly at the early stages of negotiation when the national approach is agreed or communication between the MA and urban authorities has often been limited which can lead to a lack of buy-in from urban stakeholders.

A second related issue is the challenge of coordinating a diverse and large group of actors. Coordination of design and approval of ISUD strategies is problematic, particularly in those cases where a large number of partners are involved in the approval process of the strategies. For example, in Nitra, the lack of coordination and communication between different ministries was considered an obstacle to the design process. In the context of the Maribor strategy, a lack of coordination at the central level can also lead to challenges for urban authorities in terms of multiple contacts that are responsible for different parts of the process. Similarly, the cross-sectoral nature of the strategy in Vienna is considered to have resulted in a complex coordination process.

Lastly, the governance category includes several issues in relation to the decision-making process. First, politics and negotiation can have an important impact on the development of strategies. Political will and commitment at local and central levels, as well as positive and early negotiations, were identified as shaping strategy design. In a number of instances, strong political commitment was noted. However, in others the involvement of independent experts that stood ‘above’ politics was also considered influential. For example, the Maribor strategy emphasises the importance of independent academics, not only in terms of providing analytical support but also a technical rationale for the strategy.

In some strategies, political challenges can emerge that create uncertainty and delays. In the strategies for London, the Brexit referendum caused uncertainty during the design phase. As mentioned, strategies can form the basis for political differences (Aurillac, Porto) or political changes can impact on the design process (Zagreb, Torino). The decision to implement ITI can lead to political demands from other areas to have similar arrangements (Walbrzych).

These include the following:

- Political influence can have a negative impact on the decision-making process (Maribor);
- Central-level procedures are not appropriate for the local context or are out of synch with local timelines (Debrecen, Porto, Ploiesti, Timisoara, Kaunas and Patras). For example, Aurillac pointed at the disconnection between the delegation of project identification and selection and retaining financial management, including technical assistance, which raises issues in terms of appropriate administrative resources and visibility regarding strategic management; and
- More clarity with regard to the role and responsibilities of different authorities is required (CFC pole, Six Cities, Kaunas).

6.5 Conclusion and Lessons Learned

The implementation of territorial instruments can help generate added value in various ways (Ferry et al. 2018). There is evidence of innovative policy governance approaches and administrative capacity developing at different levels and among various actors. Innovation can take three main forms: delegation of policy tasks to local levels, creation of new governance structures and strengthening of cooperative approaches. Yet, the governance of territorial instruments is creating challenges for ESIF programme managers. The effectiveness and efficiency of strategies can be undermined where existing capacities are limited. This can relate to variation in human resources available among implementing bodies and stakeholders, particularly where participation in implementing ESIF is relatively new. Designation of monitoring and control systems has been a cause of delay and drafting strategies and developing project proposals based on negotiation and consensus between partners is challenging. A difficulty for some programme authorities is the complexity associated with selecting operations, which is considered more onerous than with other ESIF operations.

Added value can also be generated through strengthened integration. Integration can be pursued at the strategic level by generating synergies between different strategic frameworks and bringing together numerous investment priorities and themes. Funding sources can also be integrated: combining different funding streams, to encourage coordinated investment in territories. Territorial integration can be pursued through a strengthened focus on functional areas or bottom-up inputs. There is also potential to develop more integrated activities at project level by combining different investments in territorial instruments and implementing a more complex and tailored set of integrated operations. Thus far, evidence indicates that integration is most notable in terms of the combination of strategic objectives in territorial instruments. Integration of funding sources and at territorial level depends strongly on governance arrangements and implementation mechanisms chosen. Most challenging is operational integration, i.e. the development of integrated activities 'on the ground'.

Thus, there are positive impacts on the involvement of local authorities in the policy process. The approaches empowered municipalities by giving them a stronger role in planning, decision-making and implementation of policies that impact on them directly. Some programme authorities have identified positive experiences and are in favour of a continuation of territorial instruments after the end of the current ESIF programme period. However, given the specific requirements for implementing these instruments and the related administrative demands, proportionality and differentiation are key concerns. This is particularly the case where the available ESIF funding is relatively low or in cases where established traditions of integrated territorial approaches to development policy limited the scope for added value.

As laid down in the TFEU territorial cohesion is a key treaty objective for the EU. The introduction of an integrated policy approach in the form of ISUD strategies, and in particular integrated territorial investment tool, have the potential to promote territorial cohesion policy goals since they contribute to increased levels of territorial efficiency and sustainability. If implemented in an appropriate manner they can reduce territorial disparities within and between urban areas in the EU. The initial evidence seems to suggest that at least in some cases the introduction of these approaches has had a significant impact on the territorial management and planning of urban policy in Member States. However, the overall impact is likely to be varied and will only become apparent over time.

References

Bache I (2010) Building multi-level governance in Southeast Europe? Southeast Eur Black Sea Stud 10(1):111–122

Bachtler J, Mendez C, Oraže H (2014) From conditionality to Europeanization in central and Eastern Europe: administrative performance and capacity in Cohesion Policy. Eur Plan Stud 22(4):735–757

Bachtler J, Polverari L, McMaster I (2009) The 'added value' of Cohesion Policy (2000–06) in the EU15; comparative report, Ex post evaluation of Cohesion Policy programmes 2000–2006 co-financed by the ERDF (objective 1 and 2)

Barca F (2009) An agenda for a reformed Cohesion Policy, a place-based approach to meeting European Union challenges and expectations. DG Regio, Brussels

Böhme K, Holstein F, Toptsidou M, Zillmer S (2017) Research for REGI committee—gold-plating in the European structural and investment funds. Available at http://www.europarl.europa.eu/RegData/etudes/STUD/2017/585906/IPOL_STU(2017)585906_EN.pdf

Dotti NF (ed) (2016) Learning from implementation and evaluation of the EU Cohesion Policy: lessons from a research-policy dialogue. RSA Research Network on Cohesion Policy, Brussels

EC (2011) Territorial Agenda of the European Union 2020—towards an inclusive, smart and sustainable Europe of diverse regions. Informal ministerial meeting of ministers responsible for spatial planning and territorial development, Gödöllő, Hungary, 19 May 2011

EC (2014) Guidance sustainable urban development. Available on http://ec.europa.eu/regional_policy/sources/docgener/informat/2014/guidance_sustainable_urban_development_en.pdf

EC (2016) Strategic plan 2016–2020, DG regional and Urban Policy, May 2016, Brussels

Farole T, Rodríguez-Pose A, Storper M (2009) Cohesion Policy in the European Union: growth, geography, institutions, working paper for the barca report. European Commission, Brussels
Ferry M, McMaster I, van der Zwet A (2017) Methodology for measuring effectiveness territorial provisions, Study conducted for the European Commission (DG Regio)
Ferry M, Kah S, Bachtler J (2018) Integrated territorial development: new instruments—new results?, IQ-net thematic paper 42(2), European Policies research centre delft
Mairate A (2006) The 'added value' of European Union Cohesion policy. Reg Stud 40(2):167–177
McCann P, Ortega-Argilés R (2016) Smart specialisation, entrepreneurship and SMEs: issues and challenges for a results-oriented EU regional policy. Small Bus Econ 46:537. https://doi.org/10.1007/s11187-016-9707-z
OECD (2009a) How regions grow. Organisation for Economic Growth and Development, Paris
OECD (2009b) Regions matter: economic recovery, innovation and sustainable growth. Organisation for Economic Growth and Development, Paris
Polverari L, Ferry M, Bachtler J (2017) the structural funds as 'agents of change': new forms of learning and implementation, IQ-net thematic paper 40(2), European policies research centre, University of Strathclyde, Glasgow
Tosics I (2015) Integrated territorial investment—a missed opportunity? paper presented at challenges for the new Cohesion Policy 2014–2020, 2nd Joint EU Cohesion Policy conference, Riga, 2–4 February 2015
van der Zwet A, Miller S, Gross F (2014) A first stock take: integrated Territorial approaches in Cohesion Policy 2014–20. IQ-net thematic paper 35(2), European policies research centre, University of Strathclyde, Glasgow
van der Zwet A, Mendez C (2015) Towards a Europe of the localities? integrated place-based approaches in Cohesion policy in 2014–20 and beyond. Paper presented at challenges for the new Cohesion Policy 2014–2020, 2nd joint EU Cohesion Policy conference, Riga, 2–4 Feb 2015
van der Zwet A, Bachtler J, Ferry M, McMaster I, Miller S (2017) Integrated territorial and urban strategies: how are ESIF adding value in 2014–2020? Final report, directorate-general for regional and Urban Policy, European Commission
van der Zwet A, Bachtler J (2018) New implementation mechanisms for integrated development strategies in ESIF. European struct investment funds J 6(1):3–12

Arno van der Zwet is a Lecturer of Politics and Public Policy at the School of Media, Culture and Society at the University of the West of Scotland and a Senior Research Associate at the European Policies Research Centre at the University of Strathclyde. He was responsible for the delivery of a major study for the European Commission, Directorate General Regio, regarding the design, implementation and management of integrated place-based strategies in European Structural and Investment Funds in the 2014–2020 period. His research interests include regional and urban development, multi-level governance, place-based policymaking, territorial politics and identity.

Martin Ferry is a Senior Research Fellow at the European Policies Research Centre (EPRC) at the University of Strathclyde in Glasgow. He specialises in research on the management and implementation of Cohesion policy and regional development policies in the United Kingdom and Central and Eastern Europe, particularly in Poland, the governance of EU Cohesion policy and pre-accession support and the evaluation of EU programmes.

Part III
Territorial Cohesion and Urban Development Policies

Chapter 7
The Urban Dimension of EU Cohesion Policy

Giancarlo Cotella

Abstract More than two thirds of EU citizens live in urban areas and cities are easily identifiable as Europe's core hubs for the promotion of territorial cohesion and overall economic growth. Whereas urban planning issues strictly remain in the hands of the Member states, since at least 30 years the EU has developed guidelines and initiatives with a more or less explicit territorial focus, among which the urban dimension had grown momentum through time. Since the publication of the Green paper on the Urban environment, a number of instruments directly targeting Member States' cities have been introduced and the urban dimension has finally entered EU mainstream cohesion policy. The chapter discusses how the lack of urban planning competences notwithstanding, the EU cohesion policy progressively developed an urban dimension. Moreover, it explores the domestic impact of the latter, also in relation to the institutional innovations that its implementation brought along with it.

Keyword EU cohesion policy · Territorial cohesion · EU urban agenda Europeanization · Spatial planning

7.1 Introduction

Cities host more than two thirds of the European Union (EU) population and constitute Europe's main development engine, playing a crucial role for both the promotion of the overall economic growth of the Union as well as in the pursuance of the objective of economic, social and territorial cohesion (CEC 2011). As a consequence, over the last 30 years, the EU has progressively developed a broad range of initiatives with regard to urban policy and urban matters more generally. Overall, it is possible to

---

G. Cotella (✉)
Interuniversity Department of Regional and Urban Studies and Planning, Politecnico di Torino, Turin, Italy
e-mail: giancarlo.cotella@polito.it

E. Medeiros (ed.), *Territorial Cohesion*, The Urban Book Series,
https://doi.org/10.1007/978-3-030-03386-6_7

analytically identify four chronological periods that characterise the development and consolidation of an EU interest in urban development matters:

(i) A first phase where the EU, after having seen its claims for urban development competences denied, develops a rather experimental approach to urban interventions (1989–1993);
(ii) A second stage, featuring more concrete conceptualizations and initiatives, that however still maintain an episodic flavour (1994–2006);
(iii) A third period, where the EU operates a first attempt to turn urban development into a mainstream element of its Cohesion Policy (2007–2013);
(iv) A fourth phase where, building on the lessons learnt from the previous programming periods, the EU mainstream approach to urban development matters is fine-tuned and provided with innovative, supporting tools (2014–2020).

In each of the four phases, the European Commission and the Member States have interacted in a cooperative-competitive game. On the one hand, the European Commission started to tackle urban issues both conceptually as well as through experimental actions. On the other hand, the member states undertook an intergovernmental cooperation process within the field of urban development, aiming at influencing the action of the Commission while reaffirming their sovereignty on the matter at the same time. This interaction has resulted in the incremental conceptualizations of the EU integrated sustainable urban development approach and of related funding instruments and actions.

Aiming at shedding light on this process, the chapter describes the development of the urban dimension of EU Cohesion Policy and of the rationale behind the latter, reflecting on the impact it had on domestic territorial governance and spatial planning. It does so drawing on the wide body of literature and official documents that have been produced on the matter since the beginning of the 1990s, as well as on part of the results of the recently concluded research project ESPON COMPASS (ESPON and TU Delft 2018; ESPON and Politecnico di Torino 2018)[1]. After this short introduction, the second section shortly presents the evolving paradigms of development that had characterised European cities starting from the 1970s, and that constitute the main scope of the EU urban policy. Section three presents the progressive development and consolidation of the mentioned EU urban policy, following the four phases introduced above and focusing on the evolution of instruments and concepts. Section four then provides some evidence of the actual impact that thirty years of EU interventions, concepts and ideas the field of urban development exerted on Member States' territorial governance and spatial planning systems. A conclusive section rounds off the contribution, reflecting upon the described issues and speculating on future development perspectives.

---

[1] The research project COMPASS—Comparative analysis of territorial governance and spatial planning systems in Europe has been commissioned by the ESPON EGTC to a research consortium led by TU Delft and composed by Nordregio, Politecnico di Torino, Polish Academy of Science, University College Dublin, Hungarian Academy of Science, Spatial Foresight, *Akademie* für Raumforschung und Landesplanung. The project lasted for the period 2016–2018. Additional information are available at: https://www.espon.eu/planning-systems.

7.2 New Paradigms of Development for European Cities

The rise of a supranational urban policy is not a direct outcome of the EU political will, rather the result of the progressive emergence of a number of more or less coherent response to the development challenges that the advent of globalization had brought with it throughout the second half of the XX century (Le Galès 2002, 2011; Matznetter and Musil 2012). As a matter of fact, since the 1970s, policy-makers and academics across Europe have become increasingly concerned with the processes of urban change and restructuring taking place in European cities. These processes had their origins with the rise of globalization trends and in the consequent decline of traditional manufacturing industries and the parallel rise of the service sector (Van Den Berg et al. 2007).

In many western and northern European countries unemployment rose dramatically. Most cities developed new forms of service-based employment, resulting in a rather modest number of well-paid positions as well as larger numbers of low-paid and insecure jobs. Many young people, entering the labour market for the first time, found themselves unable to secure permanent, reasonably well-paid jobs. These challenges took a somewhat different form in many southern European cities, partly as a result of their delayed and less intensive industrialisation,[2] but also of the different economic, social and cultural structures that through time had allowed for the development of economic networks composed by small and medium-sized enterprises, often family-owned and located outside of major cities. On their hand, many cities in the new Member States, while experiencing similar problems, also faced distinct challenges related to their communist legacy (e.g. infrastructural deficiencies, environmental problems and the legacy of mass-housing estates) (Van Kempen et al. 2005).

In many European cities, these developments found concrete expression in the growing levels of unemployment, social polarisation and segregation. In many ways, urban areas became more segmented and less cohesive. While these processes have affected the society as a whole, their impact is most clearly apparent in urban areas where whole districts experienced major large scale job loses, social tensions, minorities segregation, etc. Many run-down inner-city districts and peripheral housing estates developed large concentrations of unemployed, ethnic minorities and other vulnerable groups dependent on social assistance. Overall, these episodes of social exclusion started to be seen as a major threat to the viability of the European model of society (Faludi 2007).

These changes reflected, at the same time, a growing emphasis on economic activities based on innovation and skills that started to be considered of higher significance than cost related factors. Growth and innovation started to be based on new sectors such as the high-technology industry, neo-artisanal manufacturing, business and financial services as well as cultural and creative industries (Trip 2007). Quality of labour, quality of life, quality of environment and other 'soft factors' progressively became crucial for local development policies, leading to the conceptualization of

---

[2] Although cities such as Milan, Turin and Barcelona underwent significant industrial restructuring similar to northern cities.

new paradigms of development—'knowledge-based economy', 'creative cities', 'smart cities'—that increasingly rely on innovation and creativity as the factors stimulating development (Neef 1998; Landry 1999 ; Florida 2002; Cocchia 2014).

Another widely identified trend concerns the fundamental alteration of the spatial perspective in which cities compete with each other, as a consequence of the redefinition of spatial relations, both in the political and economic dimension. The rescaling of economies and social relations has opened an important window of opportunities for cities and regions, which have ceased to function only within their domestic systems and have increasingly become integrated into wider international flows (Brenner 2004; Swyngedouw 2004). At the same time, it also implied that development phenomena started to go beyond formal, administrative boundaries, requiring to be tackled in softer, functional levels (Haughton et al. 2011). This is especially important in the case of main development poles which gained importance in the globalized economy, and consolidated as main world players to some extent independent from their countries. Additionally, such phenomenon also become significant for the polycentric structure of many city regions where a number of towns and cities working together may consolidate into wider global flows of capital, goods and knowledge.

Finally, also the processes of policy formulation and implementation had progressively changed through time, reflecting the debate on the change from government to governance (Rhodes 1997; John 2001). The politics of urban development changed from hierarchically structured public relations to fuzzier, more inclusive public-private actions, including the private sector, social organizations, and community groups. Participatory mechanisms emerged, aiming at allowing the consultation and involvement of a broad range of stakeholders in the development of policy goals as well as in the implementation of the agreed policies.

7.3 The Progressive Consolidation of an EU Urban Agenda

It is mostly as a response to the changes introduced above, that the EU and the Member States had started to pay increasing attention to urban development issues. On the one hand, urban areas had progressively proved to be the place where the main economic and social challenges concentrate; on the other hand, they have gained recognition as the engines of economic growth for Europe, hence deserving political attention.

This occurred in parallel to, and as a consequence of, the raise of EU claims on territorial development matters, that culminated in the second half of the 1980s. In more detail, the progression of the European Economic Integration was paralleled by a growing urgency for the development of some institutional preconditions that would have allowed the EU to limit the unbalancing trends that could have resulted from it. Despite the reluctance of some Member States, the economic and social cohesion objective was included in the Single European Act in 1986, affirming the need for a supranational action on territorial development as the political condition for integration (Williams 1996; Dühr et al. 2010).

It is within this framework that the European Commission, and in particular DG REGIO, also building on the momentum generated by the subsequent reform of the structural funds in 1988, for the first time focused its radar on urban development matters. A chronological overview of how this process led to the progressive consolidation of an explicit urban dimension for EU Cohesion Policy is provided below.

7.3.1 Denied Competence and First Experimentations

The process started under rather ambitious auspices, as already during the drafting of the Maastricht Treaty there were attempts to provide the European Commission with a formal competence in the field of urban policy. The rationale for such action was clearly stated in the Green paper on the urban environment (CEC 1990), the first EU guidance document focusing on urban matters. The document, argued that "the strict zoning policies of the past decades which have led to the separation of land use and the subsequent development of residential suburbs" were not anymore adequate to deal with the evolving urban development dynamics, in so doing calling for "a fundamental review of the principles on which town-planning practice has been based", and implicitly proposing the European Commission as an active player in the process (ibidem: 40).

However, the initial optimism was wiped away by the Member States that rejected this proposal, drawing on the principle of subsidiarity. For this reason, European urban policy began to develop within the existing EU regional policy. More particularly, despite the lack of formal competences, to tackle the urban issues, the European Commission made use of Chap. 9 of the European Regional Development Fund (ERDF) Regulation, which allowed for the development of 'studies or pilot schemes concerning regional development at Community level'. As a consequence of this process, the Urban Pilot Programme was launched in 1989.

The subsidiarity principle behind which the Member States had defended their sovereignty on urban matters implied that urban related issues were to be carried out by cities themselves. Due to this reason, the Programme foresaw the use of EU ERDF funds for financing projects that were directly implemented by cities at the local level, aimed at supporting innovation in urban regeneration and planning within the framework of the broader Community policy for promoting economic and social cohesion.[3]

In the first phase (1989–1993) of the Programme, 33 Urban Pilot Projects in 11 Member States were implemented. As it was considered a success, by both the Commission and the Member states, it was then relaunched for a second round (1997–1999), leading to the implementation of 26 projects in 14 Member States. The implemented strategies combined hard infrastructure projects with social and

[3]Whereas their implementation was fully handled locally, the Commission could exert an influence on both the aims to be pursued, as well as on the governance of implementation.

economic development support, aiming at developing an 'integrated approach' to urban development and regeneration.

In parallel to the Urban Pilot Programme, another important initiative was launched in 1989 under the auspices of ERDF Chap. 9. The birth of the RECITE Programme (Regions and Cities of Europe) constituted the first step of the European Commission to promote the exchange of experiences through bottom-up networks of European cities and regions. Some of the networks developed under the RECITE initiative (e.g. Eurocities and Quartiers en crise), and those that will follow up (URBACT) had since then played a crucial role in fostering knowledge exchange and policy transfer between cities and territories, at the same time contributing to the incremental diffusion of the EU urban development concepts and ideas.[4]

7.3.2 The Consolidation of the URBAN Approach and the Birth of the Intergovernmental Cooperation

The positive reception of the Urban Pilot Programme inspired the European Commission to continue on the undertaken path, also thanks to the confidence granted by the numerous requests to become more involved in urban issues, coming from both individual cities and urban networks such as Eurocities. In 1994, a new action to support urban issues was launched, under the framework of the newly instituted Community Initiatives, specific financial instruments of the EU structural policy which were designed to find common solutions to specific widespread problems.

The URBAN Community Initiative is considered to be one of the most successful policy tools ever developed by the EU to promote integrated urban development (Carpenter 2006). The initiative aimed at developing and implementing innovative strategies for the regeneration of disadvantaged urban districts, and to spread innovative socio-economic urban development models. Each of the 118 URBAN project city, selected on the basis of calls for proposal that collectively accounted for around €953 million of EU funding, was responsible for implementing its own revitalisation programme, following an integrated urban development approach, that took into account different aspects of city life (e.g. infrastructure modernisation, revitalization of the housing stock, unemployment reduction through the creation of locally-based jobs, inclusion of disadvantaged groups, assistance for small and medium-sized enterprises, support for culture, sports and recreation in cities etc.).

At the end of the programming period, once again encouraged by the positive results, the Commission decided to relaunch this EU Community Initiative for the 2000–2006 programming period, now designated as URBAN II, which was granted around €750 million, divided across 70 different cities. The aim of both programming rounds was to contribute to the positive improvement of neighbourhoods in crisis and to develop a legacy of longer term change. Both rounds were characterised

---

[4]For a comprehensive review of policy transfer initiatives and mechanisms in Europe see Cotella et al. (2015) and Cotella (2015).

by inter-sectoral coordination of activities, concentration of funds in selected intervention areas, strong horizontal coordination of urban regeneration measures and involvement of local communities and stakeholders.[5]

In parallel to the introduction of the URBAN Initiative, and also building on the lessons learnt from the latter, the Commission continued to develop and argument its approach on urban development matters through a series of communications and position papers. Particularly relevant in this process are two documents published one after the other in 1997 and 1998: *Towards an Urban Agenda in the European Union* (CEC 1997) and *Sustainable Urban Development in the European Union: A Framework for Action* (CEC 1998). Both documents acknowledge the importance of the urban dimension in Community policies, but, in the light of previous failed attempts to gain competence on the matter, explicitly declare that this wouldn't require the attribution of additional powers.

More precisely, in the report *Towards and Urban Agenda in the European Union*, the Commission argues for an engagement of "all levels—which start from the district level to the conurbation level, up to the European urban system—within a framework of interlinking relationships and shared responsibility and achieve better policy integration" (CEC 1997: 14). On a similar line, in *Sustaining the Urban development of the European Union* report, it adds that "Urban governance can be improved by better vertical integration of activities of different levels of government and better horizontal integration within and between various organizations at the local level and involvement of stakeholders and citizens in urban policies" (CEC 1998: 19).

The debate continued in the 2000s, with the Third Report on Economic and Social Cohesion (CEC 2004) that reflected on the importance of the developing EU urban policy for the future development of cities. Cities' role as engines of regional development and attractiveness had become increasingly explicit, together with the role they could play in the promotion of economic, social and territorial cohesion (c.f. CEC 2004 and 2005). In this concern, the Communication *Cohesion policy and cities: the urban contribution to growth and jobs in the regions* (CEC 2006) constitutes a significant statement of intent, affirming that cities should remain at the top of the EU policy agenda, and paving the way for the development of a mainstream urban dimension of EU Cohesion Policy.

On their hand, the Member States did not abandon the playfield to the hegemony of the European Commission. Aiming at both preventing future competence claims as well as at proactively contributing to shape the debate according to their own domestic priorities, the Ministries responsible for spatial planning, urban affairs and regional policy launched a cooperation process in the area of urban development, as a consequence of one of the points of the ESDP Action Programme that had been approved in Tampere in 1999 (CEC 1999; FI Presidency 1999). Short after, the resulting informal intergovernmental Urban Development Group (UDG) developed a proposal for a *Multi-annual programme of cooperation in urban affairs within the European Union*, which was endorsed by the Ministries in Lille in November 2000,

[5]These principles constitute the success factors of the URBAN Initiative, as underlined the Programme Ex-post evaluation (ECOTEC 2010).

de facto giving birth to the EU Urban Agenda, a common permanent framework of reference within which, whilst recognising differences between countries and differing priorities within them, Member States could work together to develop a common approach.

Building on these developments, a series of interim resolutions and documents were then approved (NL Presidency 2004; UK Presidency 2005), eventually leading to what is considered the main milestone in the intergovernmental debate on EU Urban policy, i.e. the Leipzig Charter on Sustainable European Cities (DE Presidency 2007). Aiming at creating "a foundation for a new urban policy in Europe" (ibidem: 1), the document delivers two key policy messages: on the one hand, it argues that integrated urban development should be pursued throughout Europe, through an appropriate framework to be established on a national and EU level; secondly, it highlights that deprived urban neighbourhoods must increasingly receive political attention within the scope of an integrated urban development policy. With both the Commission and the Member States that were consolidating converging view on urban development matters, and under the auspices of the success of two round of the URBAN Community Initiative, time were ripe for the introduction of a mainstream urban dimension to EU Cohesion Policy.

7.3.3 First Attempts of Mainstreaming

The mainstreaming of the EU urban approach into the regulatory framework for the programming period 2007–2013, constituted an answer to the critic that, despite the very positive results, only a handful of selected cities had benefited from the URBAN initiative (Fig. 7.1). In this light, it aimed at enabling all European cities to promote integrated urban development with the help of EU financial support. This chapter of the EU Regulation 1080/2006 introduced the possibility of supporting the development of participative, integrated and sustainable strategies in urban areas through the use of the ERDF, the European Social Fund and the Cohesion Fund. Moreover, a number of joint financial initiatives were introduced to support urban development, such as JASPERS, JEREMIE, JASMINE and JESSICA (see CEC 2009: 36–37).

Among them, the most relevant is JESSICA (Joint European Support for Sustainable Investment in City Areas); developed as a cooperation tool between the European Commission, the European Investment Bank and the Council of Europe Development Bank, it aimed at supporting sustainable urban development and regeneration through financial engineering mechanisms. The idea was to enable Member States to use part of their EU funds to make repayable investments in urban projects supporting long-term sustainable urban renewal in the form of equity, loans or guarantees. This required the creation of a strategic framework and a public-private partnership for the delivery of a range of projects through an Urban Development Fund (for more detail see URBACT 2012).

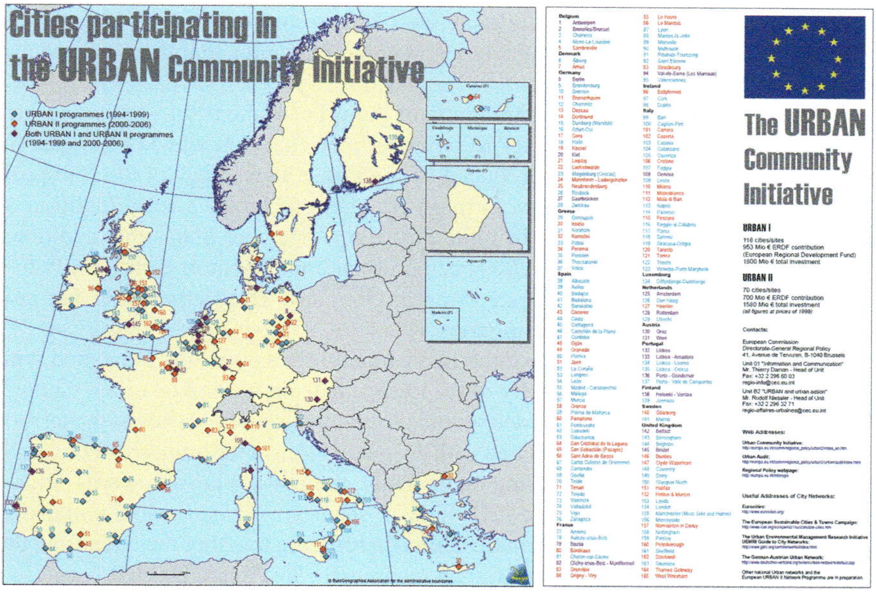

Fig. 7.1 Cities that participated in the Urban Community Initiative (*Source* CEC 2003: 9)

Whereas JESSICA was intended to "provide new opportunities for the Managing Authorities responsible for the current generation of Cohesion Policy programmes" (ibidem: 28), it however did not produce the expected results. Guidance from the European Commission and associated regulations lacked clarity and were often incompatible with national legislations. Moreover, JESSICA required a change in mentality in Managing Authorities and cities that were more familiar with using traditional grants, rather than repayable loans; similarly, it required knowledge and skills not often available within many organisations. As a result (and partly also as a consequence of the financial crisis), JESSICA suffered from underuse, despite a number of attempts of the Commission and the European Investment Bank to support regions and cities in the process (Atkinson 2014).[6]

Despite the good intentions underpinning the new approach, the Regulation did not include any incentive and/or obligation related to the creation of integrated urban development strategies to be supported under Cohesion Policy. As a consequence, in most cases, activities under EU Cohesion Policy oriented at cities were implemented in a sectoral manner, as a direct consequence of the domestic Operational Programmes' systems and associated priorities. Moreover, the possibility of delegating the responsibility for project selection and implementation to the local level was not taken into consideration and projects were normally selected at the central

[6]Only towards the end of the programming period countries and regions were beginning to use the instrument, with around 20 JESSICA holding funds that had been established in Poland, Spain, Portugal, Bulgaria, the UK, the Netherlands and Greece.

or regional level without sufficient analysis. Cities were often forced to apply for funds in the same way as all beneficiaries, with no guarantee that they will receive funds for all planned projects.[7]

Despite the fact that the mainstreaming of the urban dimension was not as successful as foreseen, it has however to be underlined that this was not the case in every EU country. A document produced by the Polish Presidency of the European Council (2011: 21–24) show how, in those countries featuring a solid tradition of national urban policies, but also in some of the new Member States that could not take part to the URBAN season, the specific urban component of Cohesion Policy was developed more thoroughly (e.g. France, Germany, The Netherlands, but also Poland, Hungary, Romania and Bulgaria).

The focus on the functioning of the mainstream urban approach coincided with a reduced conceptualization activity of the European Commission on urban matters. This field was progressively included within the conceptual boundaries of the territorial cohesion debate that was launched with the publication of the homonymous Green Paper in 2008 (CEC 2008). Building on the latter, there was a move towards a place based approach referring to a restricted range of 'special urban and spatial initiatives' and the development of a more generic approach bringing together the territorial, the social and the economic dimensions in an integrated way (Barca 2009: 93).

On its hand, the Urban Development Group continued to further develop the approach brought forward by the Leipzig Charter (FR Presidency 2008a, b). In the conclusions of the European Summit of the Regions and Cities held in Prague in 2009, the Czech Presidency called for "a strong regional policy [...] which incorporates as a crucial challenge an urban policy for the sustainable development of European cities" (CZ Presidency 2009: 16). Similar themes were also emphasised in the Toledo Declaration (ES Presidency 2010).

In the conclusions of the Multilevel Urban Governance Conference in Liege, the three Presidencies of Spain, Belgium and Hungary called upon the Member States, among others, to develop new instruments for urban multilevel coordination including all levels of government, in order to achieve the objectives of the Europe 2020 strategy (CEC 2010), the Leipzig Charter and the Toledo Declaration. Such invitation was followed up by the Polish Presidency that explicitly addressed the urban dimension of Cohesion Policy through a series of recommendations that would have then strongly influenced the 2014–2020 programming period (PL Presidency 2011).

[7]this negated the intention to achieve greater added value by implementing 'strategic and integrated policy bundles', so that individual policies would complement and reinforce one another (Atkinson 2014).

7.3.4 Fine-Tuning the Mainstream Approach. New Tools for Action

The increasing emphasis on territorial development and cohesion that was emerging within the EU discourse, also as a consequence of the inclusion of territorial cohesion in the European treaties (see also Zaucha and Świątek 2013), contributed to frame urban interventions within a wider policy context where the primary aim concerns achieving 'territorial balanced and harmonious development' across the European space (CEC 2004, 2008), while at the same time improving Europe's competitiveness. Such lied at the foundations of the Europe 2020 strategy where the emphasis on achieving smart, sustainable and inclusive growth is framed by the need to regain competitiveness (CEC 2010: 8–9). Whereas this clearly has significant implications for cities and the role they can play in achieving the above, it is worth noting that Europe 2020 has very little to say about the role of urban areas, although it does place considerable emphasis on achieving economic, social and territorial cohesion, particularly under the inclusive growth objective. As a consequence, the 2014–2020 Cohesion Policy phase proposes the 'territorial dimension' as an overarching framework within which the 'urban dimension' had to find room.

The Commission has developed a Common Strategic Framework (CSF) in order to achieve enhanced coordination between the different funds (CEC 2012: 3). At the same time, it also provided a number of general guidelines on how to use Cohesion Policy to address urban development matters. The new framework implies that a minimum of 5% of a Member State's ERDF funds should be allocated to support integrated actions for sustainable urban development. In this light, each Partnership Contract was required to include a list of cities where integrated actions for sustainable urban development are to be implemented, along with an indicative annual allocation for these actions. At the same time, Operational Programmes were allowed to include the list of cities where integrated actions for sustainable urban development were to be implemented.[8] The allocated funds could be used to tackle a broad range of economic, environmental, demographic and social issues, considered to be relevant for a particular city. Urban development has to be pursued through the characteristic EU integrated approach, addressing the specific needs of areas affected by poverty and of target groups at the highest risk of discrimination and exclusion.

Interventions allows for a good degree of flexibility, as they can be used in relation to a whole city, peri-urban areas or deprived neighbourhoods, depending on the particular challenges each city faces. Moreover, they can focus on urban-rural linkages suggesting a focus on wider functional urban regions. The intention is that whatever actions are undertaken should be strategic in nature, related to a clear logic of intervention that identifies needs, challenges and potentials. Above all the intention is

[8] However, this raised questions concerning cities selection, not only in terms of the criteria used but also whether or not cities were involved in the selection process. Most often the cities selected overlaps with key development nodes in national policy, on the one hand reinforcing national development logics, on the other hand further disadvantaging those cities that were excluded from the latter (Atkinson 2014).

that whatever is done is carried out in an integrated manner rather than in an ad hoc and disaggregated fashion as it was too often the case in the past (Atkinson 2014).

Moreover, the 2014–2020 Cohesion Policy saw also the introduction of a new set of instruments aiming at encouraging Member States and Managing Authorities to adopt a more integrated and territorially focused urban approach: the Integrated Territorial Investments (ITI) and the Community-Led Local Development (CLLD). In more detail, ITIs allows for bringing together funding from several priority axes of one or more Operational Programmes funded by the ESF and EDRF, in order to develop and implement multi-dimensional and cross-sectoral actions. Operational Programmes should identify the planned ITIs and set out an indicative financial allocation for each of them. Each ITI should then designate its target territory and integrated territorial development strategy, specify the set of actions to be implemented and set up an appropriate governance arrangement. They can focus on urban areas or other functional territories, and are implemented by a Managing Authority, a city or another public body (Tosics 2017).

On their hand. CLLDs are flexible tools for the promotion of regional and/or local development. Member States can decide to implement CLLD in selected areas/territories (e.g. urban, rural, cross-border) whose population can range from 10,000 to 150,000. The role of CLLD has to be clarified in the Partnership Contract as part of a broader integrated approach to territorial development at regional and sub-regional level in different types of territories. Overall, CLLDs can serve a range of different situations: smaller areas within cities (e.g. deprived neighbourhoods, historic centres), small and medium town and their hinterlands, or areas characterised by the presence of particularly vulnerable social groups. The approach broadly resembles that of the very successful LEADER Community Initiative, requiring setting up a local action group to develop and deliver a local development strategy that can run for up to seven years (Pollermann et al. 2014). This group is supposed to take the form of a partnership between the public and private sectors and the civil society sectors. The underlying idea is that local communities are empowered to identify the challenges their area faces and develop innovative solutions from bottom-up. Moreover, local people and businesses in the designated area should be the primary beneficiaries of the actions, so that benefits remain locally and support endogenous development.

Overall, one could argue that the opportunities for promoting integrated territorial development interventions within the framework of EU Cohesion Policy have increased in the 2014–2020 programming period. Nevertheless, the high degree of freedom provided to the Member States, in seizing these opportunities, had partially prevented this from happening. Despite acknowledging the difficulties encountered in the previous programming period in persuading Managing Authorities and cities to use new instruments for sustainable urban development, it appears that such indications were not taken into relevant account by the Commission.

The role of each single Member State in interpreting the CSF still remains very relevant, this leading to a highly differentiated application landscape. Such heterogeneity is reflected in the Partnership Contracts responsible, among others, for defining the key thematic objectives, the list of Operative programmes, their financial allocation and the implementation arrangements (Pucher et al. 2012). Arguably, there is a greater influence in those Member States where the importance of EU funds is greater (e.g. transition and less-developed regions). However, even here, the actual impact depended on how national governments drew up the Partnership Contracts and decided to address the objectives of Europe 2020 in their particular context.

7.4 The Impact of EU Urban Policy on Domestic Territorial Governance

The described development and consolidation of an EU Urban Agenda, and the instruments that progressively were implemented in the framework of the latter, contributed to promote sustainable urban development in numerous European cities. However, as it happens for the implementations of other EU policies, their concrete impact on the ground is often coupled to a more subtle institutional impact that manifests in terms of a progressive 'Europeanization' of domestic contexts (Olsen 2002; Featherstone and Radaelli 2003; Radaelli 2004; Böhme and Waterhout 2008; Stead and Cotella 2011; Cotella and Janin Rivolin 2015). According to the Europeanization literature, such impact is usually delivered through three different catalysts of change, namely (i) rules and regulations, (ii) financial resources and (iii) concepts and ideas.[9] Whereas the lack of competences on the matter does not allow the EU to produce any urban development regulation, the recently concluded ESPON COMPASS project had shown how the progressive development of concepts, ideas and instruments presented above indeed had a (differential) impact on the Member States' territorial governance and spatial planning systems (ESPON and TU Delft 2018).

First of all, through the introduction of recursive incentives addressed overall to the widespread application of EU standards of sustainable urban development, the EU progressively modified the cost-benefit logics of domestic actors and stimulated the variations of established domestic territorial governance practices. In particular, according to ESPON COMPASS experts, the EU Urban Policy had a strong or moderate impact in as many as 16 over 28 EU Member States, with Italy, Portugal, Hungary and Romania reporting the highest influence. When it comes to the geographical variation of such influence, most experts from 'old' Member States, and especially from the Mediterranean countries, highlight the importance of the Urban Pilot Projects, as well as of the URBAN Community Initiatives, and the loss of momentum registered after the cancellation of the latter and the mainstreaming of the urban dimension of Cohesion Policy in 2007.

Innovations related that spatial planning that were influenced by EU Urban policy include revitalisation plans and programmes that either take advantage of EU resources or mirrored EU programmes through domestic funds, as it happened in

[9]For a comprehensive conceptualization of the various channels of influences that are responsible for the Europeanization of territorial governance and spatial planning, see: Cotella and Janin Rivolin (2011, 2015).

Greece, Italy and Portugal. On the contrary, various Eastern European Member States that could not benefit from URBAN, among which Poland and Bulgaria, highlight the relevant influence delivered through JESSICA financial incentives and, during the present programming period, through ITIs on the domestic framework for promoting integrated sustainable urban development under the programming period 2007–2013. Similar evidences are identified in Germany, France and the Netherlands.

Beside the influence exerted through direct interventions, the EU may influence Member States' actors through circular processes of 'discursive integration' that lead EU concepts and ideas to alter beliefs and expectations of domestic actors (Böhme 2002, p. III; Adams et al. 2011; Cotella 2012). However, according to the ESPON COMPASS country experts, whereas all the document focussing on sustainable urban development played a crucial role in progressively refining the EU Cohesion Policy, they had a rather limited direct impact on domestic discourses, if compared to the impact of the described funding tools.

More pointedly, the progressive consolidation of an EU Urban Agenda is described as highly or moderately relevant only in few cases, although its relevance has been growing since 2000, mirroring its progressive development and consolidation. When compared to the other EU discursive arenas, the EU Urban Agenda explicitly records direct impacts locally, through the inspiration of integrated plans for urban regeneration, of inter-municipal partnerships, or sustainable urban strategies.[10] Moreover, in some countries, it influenced explicitly national, regional and local spatial plans such as sustainable urban mobility, urban renovation and social inclusion, and adaptation to climate change.

Most experts agree that the most influential document to the EU Urban Policies has been the Leipzig Charter on Sustainable European Cities. The publication of the latter has directly inspired the evolution of the EU Urban approach and the introduction of a number of innovative issues in local development strategies and plans, among which are: energy efficiency, sustainable mobility and sustainable urban development in general, city compactness and reduction of soil consumption, and heritage preservation (especially in Mediterranean and Central and Eastern European countries).

Overall, the evidence shows that EU Urban policy contributed to promoting a renewed interest in domestic urban policies and projects, and to introducing a programming approach to urban development issues, increasing the number and range of actors involved, promoting co-financing and the integration of resources. In general, it has contributed to a better governance of the urban dimension with increasing integration. By doing so, it has also fundamentally contributed to the promotion of Territorial Cohesion.

Other aspects of influence include knowledge transfer processes between cities, function area thinking, revitalisation for urban attractiveness, and citizens' participation. Interestingly, one could say that the urban initiatives and policy documents on the EU level are clearly rooted in the earlier European and parallel experiences

---

[10]Even though it is hard to say if the influence depends more on the persuasion capacity of the discourse itself or on the funding instruments for urban intervention put in place by the EU.

of the national urban policies. On the one hand, national programmes have helped create a basis for EU level documents, but at the same time EU interventions have often influenced the way national and local governments in the Member States have thought about the methods deployed to deal with the urban issues.

7.5 Conclusive Remarks and Future Perspectives

As we are approaching the end of the fifth EU programming period, characterised by a more or less explicit urban dimension, it is a good time to reflect on the past experiences, to both take stock of the progressive evolution of concepts and instruments developed through time at the EU level, as well as to reflect on the potentials for future development. This has been the aim of the present chapter that described the development and consolidation of an EU Urban Agenda, as an answer to the emergence and progressive acknowledgment of the new development paradigms that characterised EU cities, and reflected upon the implications that such agenda have had for the Member States and their territorial governance and spatial planning systems.

Overall, EU actions in urban matters grew incrementally throughout the 1990s up to the present day: the apparently positive experiences with urban initiatives and the growing acknowledgement that cities play a vital role in a balanced and competitive Europe, led the European Commission to suggest that sustainable urban development policies should be integrated into the mainstream of EU Cohesion Policy. In fact, during the 2007–13 period European cities benefited in many ways from Cohesion Policy instruments, initiatives and tools; and urban development issues have been integrated in all regional and national programmes supported by the Structural and Cohesion Funds (CEC 2009). Moreover, there has been a growing recognition that the EUs sectoral policies have important impacts on urban areas and their development and that these policies should take into account their 'spatial impact' and 'urban dimension' (CEC 2007).

Nevertheless, when put into perspective within the wider EU development context the actual outcomes of the urban development activities promoted through time by the European Commission have been relatively limited, mainly restricted to the development of relatively small scale 'innovative projects' or to the support of knowledge exchange networks. This is clearly illustrated in relation to the mainstreaming of the 'URBAN approach' in the two last programming periods. The intention was that all European cities would benefit from the lessons derived from URBAN and apply them to develop an integrated approach to urban areas.

However, the reality did not follow suite: whereas in some Member States (e.g. France, Germany the Netherlands, but also some new Member countries) the mainstreaming of the urban dimension produced encouraging results (PL Presidency 2011: 21–24), due to the high level of discretionality in the application of relevant regulations, numerous Member States had chosen not, or were not able to utilise those opportunities. As a result, in the majority of the Member States the new approach did not achieve relevant momentum, in some cases even leading to a weakening of the institutional innovation resulted from the implementation of the more episodic initiatives that had characterised the previous cycles.

The 2014–2020 programming period had the potential for a greater emphasis on the 'urban dimension'. However, the fact that the realisation of this potential mostly depended on the negotiations between the European Commission and the Member States, and on the Member States' interpretation of the Common Strategic Framework and use of the new instruments, again led to a highly differential landscape. Whereas in relation to the Partnership Contracts it is crucial that a range of national, regional and local stakeholders cooperated in the identification of the relevant priorities the promotion of a clear integrated territorial focus, the extent to which this has taken place remains somewhat uncertain. Across Member States, while local authorities have had some involvement in developing the Partnership Contracts, the level of involvement had varied considerably. The ITI and CLLD instruments are rather used in an experimental manner, with many Member States adapting existing delivery instruments to meet requirements for greater (territorial) integration. This is partly due to the somehow unclear relation between the overarching territorial objective of EU Cohesion Policy and the role that the urban dimension should play within the latter.

Be that as it may, the developments presented in the chapter led to a stable upload of urban matters on the EU Agenda and, eventually, to the consolidation of a recognizable and more or less autonomous EU Urban Agenda. However, despite this positive result, urban issues still remain relatively marginal to the 'mainstream' activities of the EU and the European Commission. Even today, conditions do not seem yet favourable for the development of a coherent, mainstream urban policy that produces relevant, widespread impacts in all the Member States, and that consequently contributes to achieving the EU Territorial Cohesion goal. Nevertheless, step by step, the European Commission is working in the direction of creating the conditions that may one day lead to it.

References

Adams N, Cotella G, Nunes RJ (2011) Territorial development, cohesion and spatial planning. Knowledge and policy development in an enlarged EU. Routledge, London

Atkinson R (2014) The urban dimension in cohesion policy: past developments and future prospects. Paper presented at a RSA workshop on 'The new cycle of the cohesion policy in 2014–2020'. Institute for European Studies, Vrije Universiteit Brussels, 24 Mar 14

Barca F (2009) An agenda for a reformed cohesion policy, a place-based approach to meeting European Union challenges and expectations. Independent report prepared at the request of Danuta Hübner, Commissioner for Regional Policy, Brussels

Böhme K (2002) Nordic echoes of European spatial planning. Nordregio, Stockholm

Böhme K, Waterhout B (2008) The Europeanization of planning. In: Faludi A (ed) European spatial research and planning. Lincoln Institute of Land Policy, Cambridge, MA, pp 225–248

Brenner N (2004) New state spaces. Urban governance and the rescaling of statehood. Oxford University Press, Oxford, New York

Carpenter J (2006) Addressing Europe's urban challenges: lessons from the EU URBAN community initiative. Urban Studies 43(2):2145–2162
CEC—Commission of the European Communities (1990) Green paper on the urban environment. Communication from the Commission to the Council and the Parliament. COM(90) 218 Final, 27 June 1990. Commission of the European Communities, Brussels
CEC—Commission of the European Communities (1997) Towards an urban agenda in the European Union, communication from the Commission, COM(97) 197, final, 06 May 97. Commission of the European Communities, Brussels
CEC—Commission of the European Communities (1998) Sustainable urban development in the European Union: a framework for action, Communication from the Commission, COM(98) 605 final, 28 Oct 98. Commission of the European Communities, Brussels
CEC—Commission of the European Communities (1999) ESDP—European spatial development perspective. Towards balanced and sustainable development of the territory of the EU, approved by the Ministers responsible for Regional/Spatial Planning of the European Union. European Communities, Luxembourg
CEC—Commission of the European Communities (2003) Partnership with the cities. The URBAN Community Initiative. Office for official publications of the European Communities, Luxembourg. Available at: http://ec.europa.eu/regional_policy/sources/docgener/presenta/cities/cities_en.pdf
CEC—Commission of the European Communities (2004) A new partnership for cohesion. Convergence, competitiveness, cooperation, Third Report on economic and social cohesion. Office for the official publications of the European Communities, Brussels. Available at: http://ec.europa.eu/regional_policy/sources/docoffic/official/reports/cohesion3/cohesion3_en.htm
CEC—Commission of the European Communities (2005) Cities and the Lisbon Agenda: assessing the performance of cities. Directorate General Regional Policy, Brussels
CEC—Commission of the European Communities (2006) Cohesion policy and cities: the urban contribution to growth and jobs in the regions, Communication from the Commission to the Council and Parliament, COM(2006) 385 final of 13 July 2006, Brussels
CEC—Commission of the European Communities (2007) The urban dimension in community policies for the period 2007–2013. Commission of the European Communities, Brussels
CEC—Commission of the European Communities (2008) Turning territorial diversity into strength. Green paper on territorial cohesion. Office for Official Publications of the European Communities, Luxembourg
CEC—Commission of the European Communities (2009) Promoting sustainable urban development in Europe. Achievements and opportunities. Commission of the European Communities, DG Regio, Brussels
CEC—Commission of the European Communities (2010) Europe 2020. A strategy for smart, sustainable and inclusive growth, COM(2010) 2020 final. Office for Official Publications of the European Communities, Luxembourg
CEC—Commission of the European Communities (2011) Cities of tomorrow. Challenges, visions, ways forward. Office for the official publications of the European Communities, Brussels
CEC—Commission of the European Communities (2012) Elements for a common strategic framework 2014 to 2020 the European Regional Development Fund the European Social Fund, the Cohesion Fund, the European Agricultural Fund for Rural Development and the European Maritime and Fisheries Fund. Commission Staff Working Document. SWD (2012) 61. Final, 14 Mar 2012. Commission of the European Communities, Brussels
Cocchia A (2014) Smart and digital city: a systematic literature review. In: Dameri R, Rosenthal-Sabroux C (eds) Smart city. Springer, Amsterdam, pp 13–43
Cotella G (2012) Central and Eastern European actors in the European spatial planning debate: time to make a difference? Europa XXI 22(2012):21–36
Cotella G (2015) Transferring territorial governance in the European Union: why, what, how and through whom? Europa XXI 26(2014):45–63
Cotella G, Janin Rivolin U (2011) Europeanization of spatial planning through discourse and practice in Italy. disP Plann Rev 186:42–53

Cotella G, Janin Rivolin U (2015) Europeizzazione del governo del territorio: un modello analitico. Territorio 73:127–134
Cotella G, Janin Rivolin U, Santangelo M (2015) Transferring good territorial governance in Europe: opportunities and barriers. In: Schmidt P, Van Well L (eds) Territorial governance across Europe: pathways, practices and prospects. Routledge, London, pp 238–253
CZ Presidency (2009) Regions and cities for Europe, Report of the European summit of regions and cities for Europe, Praha, 5–6 Mar 2009
DE Presidency (2007) Leipzig charter on sustainable European cities. Final Draft, Leipzig
Dühr S, Colomb C, Nadin V (eds) (2010) European spatial planning and territorial cooperation. Routledge, London & New York
ECOTEC (2010) Ex post evaluation of Cohesion Policy Programmes 2000–2006. Executive summary, July 2010. ECOTEC
ES Presidency (2010) Toledo informal ministerial on urban development. Declaration, Toledo, 22 June 2010
ESPON and Politecnico di Torino (2018) ESPON COMPASS—comparative analysis of territorial governance and spatial planning systems in Europe. Volume 7—Europeanization. ESPON EGTC, Luxembourg
ESPON and TU Delft (2018) ESPON COMPASS—comparative analysis of territorial governance and spatial planning systems in Europe. Final report. ESPON EGTC, Luxembourg
Faludi A (ed) (2007) Territorial cohesion and the European model of society. Lincoln Institute of Land Policy, Cambridge, MA
Featherstone K, Radaelli CM (eds) (2003) The politics of Europeanization. Oxford University Press, Oxford
FI Presidency (1999) European spatial development perspective action programme, 1999. Informal Meeting of EU Ministers Responsible for Spatial Planning and Urban/Regional Policy of the European Union. Saarijärvi, Gummerus Printing, Tampere
Florida R (2002) The rise of the creative class. Basic Books/Perseus, New York
French Presidency (2008a) Final statement by the ministers in charge of urban development. Présidence Française Union Européenne, 25 Nov 2008
French Presidency (2008b) Final appendix to the final statement by the ministers in charge of urban development for implementing the Leipzig Charter. Présidence Française Union Européenne, 25 Nov 2008
Haughton G, Allmendinger P, Counsell D, Vigar G (2011) The new spatial planning: territorial management with soft spaces and fuzzy boundaries. Routledge, London
John P (2001) Local governance in Western Europe. Sage, London
Landry C (1999) The creative city. A toolkit for urban innovators. Routledge, London
Le Galès P (2002) European cities. Social conflicts and governance. Oxford University Press, Oxford
Le Galès P (2011) Le retour des villes européennes, Sociétés urbaines, mondialisation, gouvernement et gouvernance, 2e édition augmentée d'une préface inédite, Presses de la Fondation Nationale des Sciences Politiques, Paris
Matznetter W, Musil R (2012) The European City in the age of globalisation, Belgeo [On-line]. URL: http://journals.openedition.org/belgeo/6100
Neef D (1998) The knowledge economy. Butterworth-Heinemann
NL Presidency (2004) Urban acquis. Rotterdam http://ec.europa.eu/regional_policy/urban2/pdf/112004_rotterdam_conclusion.pdf
Olsen JP (2002) The many faces of Europeanization. J Common Market Stud 40(5):921–952
Pollermann K, Raue P, Schnaut G (2014) Multi-level Governance in rural development: analysing experiences from LEADER for a Community-Led Local Development (CLLD). In: 54th congress of the European Regional Science Association: "regional development & globalisation: best practices", 26–29 Aug 2014, St. Petersburg, Russia, ERSA
PL Presidency (2011) Background report on the urban dimension of the Cohesion Policy post 2013. Report prepared at the request of the Polish EU Presidency. Polish Ministry of Regional Development, Warsaw

Pucher J, Naylon I, Resch A (2012) The partnership contracts—how to implement multilevel governance and to guarantee flexibility of Cohesion Policy. European Parliament, Policy Department B: Structural and Cohesion Policies, Brussels
Radaelli CM (2004) Europeanization: solution or problem? Eur Integr Online Papers 8(16):1–16
Rhodes R (1997) Understanding governance. Open University Press, Buckingham
Stead D, Cotella G (eds) (2011) Differential Europe, special issue. disP The Plann Rev 186(3):12–83
Swyngedouw E (2004) Globalisation or 'glocalisation'? Networks, territories and rescaling. Cam Rev Int Affairs 17(1):25–48
Tosics I (2017) Integrated territorial investment—a missed opportunity. In: Bachtler J, Berkowitz P, Hardy S, Muravska T (eds) EU cohesion policy. Reassessing performance and direction. Routledge, London/New York, pp 284–296
Trip JJ (2007) Assessing quality of place: a comparative analysis of Amsterdam and Rotterdam. J Urban Affairs 29(5):501–517
UK Presidency (2005) Conclusions of ministerial informal meeting on sustainable communities in Europe, Bristol, 6–7 Dec 2005
URBACT (2012) JESSICA 4 cities. How cities can make the most from Urban Development Funds, URBACT
Van den Berg L, Braun E, van der Meer J (eds) (2007) National policy responses to urban challenges in Europe. Ashgate, Aldershot
Van Kempen R, Vermeulen M, Baan A (eds) (2005) Urban issues and urban policies in the New EU Countries. Ashgate, Aldershot
Williams RH (1996) European Union spatial policy and planning. Paul Chapman Publishing, London
Zaucha J, Świątek D (2013) Place-based territorially sensitive and integrated approach. Polish Ministry of Regional Development, Warsaw

Giancarlo Cotella is Associate Professor of Spatial Planning at Politecnico di Torino. His research focuses on EU Territorial Governance, in particular on the comparison of the mutual influence between European Spatial Planning and the territorial governance and spatial planning systems of the Member States. In recent years, he took part in several international research projects, among which ESPON 2.3.1 ESDP, ESPON 2.3.2 Governance, ESPON FOCI, ESPON SMART-IST, ESPON TANGO, ESPON ReSSI, ESPON URRUC and ESPON COMPASS. He lectured in numerous universities around the world and published several contributions on international scientific journals and books. Since 2006, he is actively involved in the Association of European Schools of Planning.

Chapter 8
Policy Cycle of the Urban Agenda for EU and Its Effects on Territorial Cohesion

Paulo Neto, Maria Manuel Serrano and Anabela Santos

Abstract This chapter aims to analyse how, in the European Union's public policy set, policies directed to cities and urban matters have become more relevant. The importance of these policies in the European agenda is seen in Cohesion Policy, but also in the Union's other policies, with other thematic and sector focuses. The intention is also to analyse to what extent the transversal, trans-sector, multi-scale and transpolicy character—which urban questions have acquired in European policies—are contributing to the development of a new generation of public, national and regional policies, and to achieving higher levels of territorial cohesion. The pertinence of this research is reinforced by the creation of the Urban Agenda for EU, created in 2016 by the European Union, which aims to systematise and reorganise the different Union policies directed to cities and revalue their role in the process of European integration.

Keywords Urban agenda · Policy cycle · EU cohesion policy · European Union Territorial cohesion

---

P. Neto (✉)
Department of Economics, University of Évora, Public Policy Monitoring Unit (UMPP), CICS.NOVA.UÉvora and CEFAGE-UÉ, Évora, Portugal
e-mail: neto@uevora.pt

M. M. Serrano
Department of Sociology, University of Évora, Public Policy Monitoring Unit (UMPP), CICS.NOVA.UÉvora and SOCIUS-CSG/ISEG-UL, Évora, Portugal
e-mail: mariaserrano@uevora.pt

A. Santos
International Centre for Innovation, Technology and Education Studies (iCite), Solvay Brussels School of Economics and Management, Université Libre de Bruxelles, Brussels, Belgium
e-mail: asantos@ulb.ac.be

E. Medeiros (ed.), *Territorial Cohesion*, The Urban Book Series,
https://doi.org/10.1007/978-3-030-03386-6_8

8.1 Introduction

The European Commission defended that "although cities' role for economic, social and cultural development, and their potential for a more resource-efficient habitat, have long been recognised, the policy response at European and national level has been slow and piecemeal, with many but poorly integrated sectoral initiatives. A growing number of voices argue that cities need to be adequately involved in the conception and implementation of EU policies, and that EU policies need to be better adapted to the urban realities where they will be implemented" (European Commission 2014: 3).

In this sense, "despite cities' potential for driving growth, the highest unemployment rates are found in cities. With globalisation, the recent crisis and a drop in manufacturing, many cities have experienced a de-skilling of the workforce, and an increase of low-skilled service sector jobs and working poor. The share of the population at risk of poverty has increased. Many cities face a significant increase in social exclusion, segregation and polarization" (European Commission 2014: 4).

The European Commission's *Seventh report on economic, social and territorial cohesion* recognises that "despite the growing concentration of jobs in cities, the share of low work intensity households is the highest in EU-15[1] cities. The risk of poverty or social exclusion in the EU has fallen back to its pre-crisis level. In the EU-13,[2] it is even lower than before the crisis, but in the EU-15 it remains higher than before in cities, towns and suburbs. This highlights the fact that pockets of poverty exist even in relatively well-off cities" (European Commission 2017: xv).

The report *The State of European Cities 2016; Cities leading the way to a better future* from the European Commission and the United Nations Human Settlements Programme (UN-Habitat) 2016 presents European cities and the European urban system as follows: "the share of population in European cities has barely changed in the last fifty years and it is relatively low by global standards. European cities tend to be mid-sized with few cities over one million and only two over 10 million inhabitants. European cities have lower population densities than Asian cities, but are still more than twice as dense as North American cities. European cities are on average located closer to each other than cities in other parts of the world, but the closest large city is much further away. This is the outcome of Europe's dense network of mid-sized cities in general and because they tend to be less clustered around large cities" (European Commission 2016a: 32).

In the European Union, "capital regions, cities and sometimes even smaller towns (e.g. in traditional holiday zones) have stronger GDP growth, more employment opportunities than other regions and hence, attract people. Consequently, about 75% of the EU population today is concentrated in urban areas. However, important differences in the breakdown of urban population (capitals/cities/towns) are observed in

---

[1]EU-15: Austria, Belgium, Denmark, Finland, France, Germany, Greece, Spain, Ireland, Italy, Luxemburg, The Netherlands, Portugal, Sweden and The United Kingdom.

[2]EU-13: Bulgaria, Croatia, Cyprus, Czech Republic, Estonia, Hungary, Latvia, Lithuania, Malta, Poland, Romania, Slovakia and Slovenia.

the EU. Despite the projected modest increase of the total EU population, peripheral regions are expected to experience a decline in their city population, while capital regions in central Europe are likely to grow" (Lavalle et al. 2017: 3).

This chapter aims to analyse how, in the European Union's public policy set, policies directed to cities and urban matters have become more relevant. The importance of these policies in the European agenda is seen in EU Cohesion Policy, and also in the Union's other policies, with other thematic and sector focuses. The intention is also to analyse to what extent the transversal, trans-sector, multi-scale and trans-policy character—which urban questions have acquired in European policies—are contributing to the development of a new generation of public, national and regional policies, and to achieving higher levels of territorial cohesion. The pertinence of this research is reinforced by the elaboration of the Urban Agenda for EU, created in 2016 by the European Union, which aims to systematise and reorganise the different Union policies directed to cities and revalue their role in the process of European integration.

8.2 The Urban Dimension in the Process of European Integration and in Defining the Union's Public Policy Set

There are many definitions of a city. "City' can refer to an administrative unit or a certain population density. A distinction is sometimes made between towns and cities—the former are smaller (e.g. between 10,000 and 50,000 inhabitants) and the latter are larger (above 50,000 inhabitants). 'City' can also refer more generally to perceptions of an urban way of life and specific cultural or social features, as well as functional places of economic activity and exchange. 'City' may also refer to two different realities: the 'de jure city'—the administrative city—and the 'de facto city'—the larger socio-economic agglomeration. The 'de jure city' corresponds to a large extent to the historic city with its clear borders for trade and defence and a well-defined city centre" (European Commission 2011b: 1).

In *Towards an urban agenda in the European Union*,[3] the European Commission defended the need to create an urban perspective in European Union policies, where "EU should play a complementary role in addressing urban issues as it has responsibility for policies in a number of sectors which have a direct bearing on the development and quality of life in urban areas. Possibilities for adapting these policies to improve their contribution to urban development need to be more exhaustively explored" (European Commission 1997: 14).

Among the areas for further reflection in this perspective, the European Commission (1997) proposed the following: (i) the development of clear targets for improvement of the urban environment with specified timescales, and the improvement of EU sectoral policies from the viewpoint of sustainability; (ii) the development of efficient

[3]COM (97) 197 final.

access to the networks from regional and local systems, and to ensure that resources are used to produce the maximum benefit in terms of environmental, employment and industrial objectives; (iii) the reinforcement of intermodal freight and passenger transport, both for facilitating access to the city and mobility within the city; (iv) the targeting of RTD activities on the main problems facing the cities of tomorrow, namely, integrated transport, energy, sustainable construction technology, information networks, technology for the protection of cultural heritage, urban sustainable development, environmental technologies and new urban vehicles; (v) telecommunications policies, including Universal Service obligations, to ensure the earliest provision of links to the information highway involving depressed urban neighbourhoods and smaller urban areas; (vi) the strengthening of the commercial function of cities and neighbourhoods and their role in the development of tourism; (vii) the issues of migration, police and judicial cooperation; (viii) the fight against social and economic exclusion, which is an explicit goal of the European employment strategy, as well as the fight against racism in the framework of the 1997 European Year Against Racism; (ix) public health policy and in particular health concerns related to urban deprivation and poverty; and (x) the need for creating trust-based relationships between various actors at the local level, in order to promote local empowerment, responsibility and initiative, and to reinforce employment policies.

In the *Sustainable Urban Development In The European Union*[4] document: the European Commission presented the EU Framework for action for sustainable urban development, organised under four interdependent policy aims: (i) strengthening economic prosperity and employment in towns and cities; (ii) promoting equality, social inclusion and regeneration in urban areas; (iii) protecting and improving the urban environment: towards local and global sustainability; and (iv) contributing to good urban governance and local empowerment.

In 2005, the European Commission started a process of urban audit at the European level with the report *Cities and the Lisbon Agenda: Assessing the Performance of Cities*. One year later, it presented the report *Cohesion Policy and cities: the urban contribution to growth and jobs in the regions*[5] and explained 'why cities matter'. There was the conviction that "the European Union will pursue its objectives of growth and jobs more successfully if all regions are able to play their part [and cities] are particularly important in this context" (European Commission 2006: 4).

In 2007, through *The Urban Dimension in Community Policies for the Period 2007–2013* document, the European Commission detailed the urban dimension in Community Strategic Guidelines (CSG). The CSG specify that programmes with a focus on urban areas can take different forms: 'there are actions to promote cities as motors of regional development [...] other actions aim to promote internal cohesion inside urban areas by improving the situation of neighbourhoods in crisis, notably by rehabilitating the physical environment, redeveloping brownfield sites and preserving and developing the historical and cultural heritage, [...] other actions aim to promote a balanced, polycentric development of the European Union by developing

[4]COM (1998) 605 final.

[5]COM (2006) 385 final.

the urban networks at national and Community level. To achieve this objective, it is a matter of putting in place networks linking cities in both physical (e.g. infrastructure, information technologies, etc.) and human (e.g. promotion of cooperation, etc.) terms. Specific attention should also be paid to the urban-rural interface' (European Commission 2007: 8).

In the year 2008, the *Green Paper on Territorial Cohesion: Turning Territorial Diversity into Strength*[6] is responsible for a major new impetus in the European urban policy due to the effort to intensify the territorial dimension of economic and social cohesion.

In this sense, there seems to be growing consensus and a shared vision in the context of the European Union regarding the objective of what cities should be. This exercise of designing cities 'as they should be' contemplates, among other aspects, the following: (i) a place of advanced social progress with a high degree of social cohesion and socially balanced housing, as well as social, health and 'education for all' services; (ii) a platform for democracy, cultural dialogue and diversity; (iii) a place of green, ecological or environmental regeneration; and (iv) a place of attraction and an engine of economic growth. In this context, the key principles of what should be the future of European urban and territorial development are sustained on aspects such as: (i) being based on balanced economic growth and territorial organisation of activities, with a polycentric urban structure; (ii) building on strong metropolitan regions and other urban areas that can provide good accessibility to services of general economic interest; (iii) being characterised by a compact settlement structure with limited urban sprawl; and (iv) enjoying a high level of environmental protection and quality in and around cities (European Commission 2011b).

Atkinson argues that "over the last 20 years, the EU has been involved in a wide range of initiatives with regard to urban policy and urban matters more generally. As a result, what might be described as a 'European Urban Agenda' has emerged. Initially the EU, during an early exploratory phase, took on a more observational and 'experimental' role, but in the last decade its recommendations and activities have become more and more concrete. It has set up programmes (e.g. URBAN I and II, subsequently 'mainstreamed' in the last period), facilitated studies, research and the dissemination of good practices as well as supporting innovative projects across Europe" (Atkinson 2014: 2).

The Barca Report (2009) proposed the place-based approach as the new rationale for action for EU Cohesion Policy 2014–2020. For the author, "place-based policy is a long-term strategy aimed at tackling persistent underutilisation of potential and reducing persistent social exclusion in specific places through external interventions and multilevel governance […]. In a place-based policy, public interventions rely on local knowledge and are verifiable and submitted to scrutiny, while linkages among places are taken into account" (Barca 2009: vii).

Closely associated with this concept of the place-based approach is the objective of achieving better results in terms of territorialisation of public policies. Medeiros adds that "alongside the policy goals and related strategic execution, the territorialisation

[6]COM (2008) 616 final.

of a given policy depends on its effects in promoting territorial development and/or territorial cohesion [...] we define the territorialisation of a given policy as its capacity to: (i) encompass the main dimensions of territorial development; (ii) effectively promote territorial development or cohesion, and; (iii) target a specific territorial scale in its strategic intervention" (Medeiros 2016: 95).

Another set of authors, in the *Scenarios for Integrated Territorial Investments* document, argue that "the main focus of a place-based policy is on the effective use of each territory's potential. This guarantees long-term socio-economic benefits both for the local population and others living within a certain radius of the territory in question. The approach involves close dialogue and cooperation between institutions and actors operating at different management levels. It also combines external know-how and resources with local assets in a way that eliminates or decreases the threat of domination by any of the partners. Tailoring public policies to the needs on the ground should bring long-lasting effects to the local population. It should also help to achieve both European and national goals thanks to the proper use of locally rooted development factors like infrastructure, governance culture or climate" (De Bruijn et al. 2015: 15).

As demonstrated, EU Cohesion Policy (2014–2020) has provided new instruments relevant to the urban dimension, namely, the: (i) Integrated Territorial Investment (ITI); (ii) Integrated Sustainable Urban Development and (iii) Community-Led Local Development (CCLD). "In addition the general use (or mainstreaming) of the LEADER approach, which shares much in common with URBAN, offers enhanced encouragement for Member States and Managing Authorities to adopt a more integrated and territorially focused approach that has a significant 'bottom-up' component and allows local communities to take a leading role in the design and delivery. It is these new instruments that offer the potential for an enhanced 'urban dimension'" (Atkinson 2014: 12).

According to another reading of the subject, "the introduction of integrated place-based approaches has led to some important changes in the manner in which strategies are designed, specifically (one of the key characteristics of place-based approaches) the interaction between bottom-up local knowledge and top-down operational and analytical expertise. In terms of the bottom-up process, the early engagement of the public and other key stakeholders in the strategy development process is considered important good practice for urban and regional authorities, which encourages buy-in, ownership and collective responsibility" (Zwet et al. 2017: 85).

For the Cohesion Policy programming period 2014–2020, the development of a place-based policy, under Smart Specialisation Strategies (S3), is becoming a prerequisite for accessing EU funding. Consequently, the EU Urban Policy for this period was also designed from an integrated approach.

8.3 Evolution of Urban Policy in the Context of EU Cohesion Policy's Policy Cycle: From a Specific Public Policy to a Multiple, Transversal and Trans-Sector Public Policy

In the Declaration of Ministers Responsible for Territorial Cohesion and Urban Matters, regarding the EU Urban Agenda,[7] following the Informal Meeting of June 2015, it was decided that the support for development of the EU Urban Agenda, among other aspects, should (i) be taken forward in close partnership among Member States, the Commission, the European Investment Bank (EIB), regions, urban authorities and other concerned stakeholders, according to the principles of subsidiarity and proportionality; (ii) respect urban diversity and specificities, as well as contributing to unlocking the potential of all types of urban areas, including small- and medium-sized ones, and the most vulnerable and deprived neighbourhoods; (iii) provide an operational framework and effective instruments to horizontally improve the urban dimension in European policymaking, including better policies, territorial impact assessment, more tailor-made and place-sensitive EU funds, including EIB finance, opportunities to exchange knowledge and best practices, research and cooperation; (iv) promote cooperation among urban areas of all sizes, as well as cooperation within functional urban areas and regions, including inner areas and cross-border polycentric metropolitan regions; (v) promote the knowledge base and data on urban development issues, referring to different types of urban units; (vi) support policies, at national or regional level, that have an impact on and promote sustainable and integrated urban development and territorial cohesion, and contribute to the overall EU strategic objectives, raising awareness and fostering the coordination of sectors; and (vii) strengthen the intergovernmental cooperation on urban and territorial development, in order to provide a clear contribution to the development and implementation of the EU Urban Agenda.

The Urban Agenda is an "integrated and coordinated approach to deal with the urban dimension of EU and national policies and legislation. By focusing on concrete priority themes within dedicated Partnerships, the Urban Agenda seeks to improve the quality of life in urban areas. It is a new working method to ensure maximum utilisation of the growth potential of cities and to successfully tackle social challenges and it aims to promote cooperation between Member States, Cities, the European Commission and other stakeholders".[8]

The paradigm of how to design urban planning has changed. In fact "over recent decades, urban planning has evolved from being a merely technical discipline into something much more complex, where politicians and stakeholders want urban development to meet the needs of cities and people. Many years of practical experience have delivered valuable elements of good practice and recommendations for urban

---

[7]See https://eu2015.lv/images/news/2015_06_10_EUurbanDeclaration.pdf.

[8]See https://ec.europa.eu/info/eu-regional-and-urban-development/topics/cities-and-urban-development_en.

policy that meet the challenges in European cities. There are at least five dimensions or features that can be identified: (i) a move away from individual sectors towards wider integration within the local or regional economy; (ii) a shift from government to governance, i.e. the tendency of central governments to confer certain duties to lower levels of government, such as provinces, regions, cities, city districts and neighbourhoods (referred to as 'decentralisation'). Together with the privatisation of governmental tasks, this involves the participation of a larger number of different policy partners, organisations and individuals (governance); (iii) an increasing focus on empowering the inhabitants of cities and specific neighbourhoods; (iv) a shift from universal policies to more focused, area-based policies; (v) growing attention paid to the effectiveness of policies" (European Commission 2009: 25).

The methodological advances in addressing these themes were also evident "a common methodology for sustainable urban development has begun to take shape over the last decade and has been generated following the emergence of a European 'Acquis Urbain',[9] which builds on the experience gained while supporting integrated and sustainable urban development. This methodology is also in line with the policy principles and recommendations laid down in the Leipzig Charter on Sustainable European Cities. The 'Acquis Urbain' is based on the following cornerstones[10]: (i) the integrated and cross-sector approach of the URBAN Community Initiatives; (ii) the new instruments of urban governance, administration and management, including increased local responsibilities and strong horizontal partnerships, successfully tested by the URBAN Community Initiatives; (iii) a targeted selection of towns, cities and eligible areas and the concentration of funding; (iv) networking, benchmarking and the exchange of knowledge and know-how, building on the positive experience and results of the URBACT I Programme" (European Commission 2009: 25).

Zwet et al. (2017) argue that there has been a significant uptake of territorial strategies in the 2014–2020 EU Cohesion Policy programming period, mainly in the form of promoting sustainable urban development processes, across most Member States. Furthermore, many of the strategies are new, and the territorial provisions have encouraged innovation and adaptation in both thinking and practice. This innovation includes a more integrated approach to intervention, the implementation of strategies at different spatial scales and more collaborative models of governance.

Urban-oriented initiatives and programmes have contributed to promote "a sort of 'policy style' based on networking, cross-dissemination (URBACT), multi-actor participation (local development partnerships) and the use of management tools in order to produce comparable and evaluable urban data (Urban Audit, EUKN or ESPON). However, the most significant mechanism of influence has been the conditionality introduced by the Structural Funds, especially the European Regional Development

---

[9]See The 'Acquis URBAN'—Using Cities' Best Practices for European Cohesion Policy (Common Declaration of URBAN cities and players at the European Conference 'URBAN Future', Saarbrücken, 8–9 June 2005).

[10]COM (2006) 385 final.

Fund (ERDF) through the Community Initiative URBAN and currently the article 7 of ERDF" (Medina and Fedeli 2015: 4).

There is growing consensus that cities must work from a cross-sectoral perspective and not allow single-sector views to determine the future agenda of urban life. Horizontal and vertical coordination is necessary as cities must work together with other levels of governance and strengthen cooperation and networking with other cities in order to share investments and services that are needed on a broader territorial scale (European Commission 2011b).

Just as there is increasing recognition of the relevance of a multi-scale and trans-policy approach to urban policy, and "strategies to support urban regeneration should always be promoted with an overall and systemic vision. Articulating the various interventions planned for each of the parts of the city in order to ensure an overall coherence and systemic coordination as well as the full realisation of all multiplier effects. As well as combining, in urban regeneration processes and landscape, reshaping, strategies to safeguard and to potentiate urban identity, urban plasticity, urban temporality, image, and attractability, and to assure the economic, social and cultural dimension in the expected results of the territorial strategies implemented" (Neto 2014: 83).

Tables 8.1 and 8.2 present the current set of policy instruments for urban policy in the European Union. The Union's urban policy is now based on a range of initiatives, from funding, knowledge production and support, to cooperation and recognition of good practice, with a multisectoral and transpolicy perspective.

Assessing the effectiveness of integrated territorial and urban strategies in the future should focus on the following principles: "(i) to recognise the need for differentiation in the size and complexity of indicators sets, in line with variation in budgets and scope of strategies (…); (ii) to capture the results of territorially integrated approaches in line with the logic of the intervention. The aim is to define synthetic or integrated indicators that can assess combined actions (…); (iii) added value. In many cases, these territorial initiatives represent new or innovative ways of doing things and the reasons for introducing them include institutional and operational, as well as physical change. This highlights the need for a combination of 'hard' and 'soft' indicators" (Ferry et al. 2018: 25).

Concerning the Europe 2020 Index and the progress of EU countries, regions and cities to the 2020 targets, Dijkstra and Athanasoglou consider that: "(i) the analysis shows that cities score on average higher on the 2020 index than towns, suburbs and rural areas; (ii) the difference between the score of cities and areas outside cities can be very high, especially in Cohesion countries; (iii) non-Cohesion countries' cities do not perform better than their towns, suburbs and rural areas on the Europe 2020 index; (iv) the cities in Cohesion countries almost match the performance of cities in non-Cohesion countries; (v) in Cohesion country cities, employment rates are slightly lower and poverty or exclusion rates are slightly higher, but their tertiary education rates are equally high and the share of early school leavers is only half that of cities in non-Cohesion countries" (Dijkstra and Athanasoglou 2015: 11).

Table 8.1 European Union's city initiatives, knowledge support and funding

Cities initiatives	Knowledge for cities	Funding for cities
• Sustainable urban development • Smart cities • Working on urban development globally – New Urban Agenda[a] – Union for Mediterranean Urban Agenda[b]	• ESPON[c] • Urban Data Platform[d] • Copernicus Urban Atlas[e] • Global Human Settlement Layer (GHSL)[f] • Urban Audit[g] • Urban Europe—statistics on cities, towns and suburbs[h]	• European Fund for Strategic Investments (EFSI)[i] • European structural and investment funds • Horizon 2020[j] • LIFE[k] • Urban Innovative Actions[l] • European Investment Project Portal (EIPP)[m] • European Investment Advisory Hub – URBIS[n]

Source Authors' own elaboration based on European Commission, https://ec.europa.eu/info/eu-regional-and-urban-development/topics/cities-and-urban-development_en

[a]The New Urban Agenda was adopted by the United Nations at the Habitat III conference on Housing and Sustainable Urban Development in October 2016. See http://habitat3.org/wp-content/uploads/NUA-English.pdf

[b]The Union for Mediterranean Urban Agenda is about regional cooperation in sustainable urban development between the EU and countries on the Southern and Eastern shores of the Mediterranean. See http://ufmsecretariat.org/ufm-ministers-agree-on-a-structured-framework-for-enhanced-regional-cooperation-on-sueurinable-urban-development-in-the-euro-mediterranean-region/

[c]ESPON EGTC, European Spatial Planning Observation Network, see https://www.espon.eu/. See also ESPON CityBenchWebtool http://citybench.espon.eu/citybenchwebclient/

[d]See: http://urban.jrc.ec.europa.eu/?ind=popden&ru=fua&s=0&c=1&m=0&f=1&p=0&swLat=35.88905007936091&swLng=-40.693359375&neLat=59.712097173322924&neLng=62.40234375

[e]See https://land.copernicus.eu/local/urban-atlas

[f]See http://ghsl.jrc.ec.europa.eu/

[g]See http://ec.europa.eu/regional_policy/en/policy/themes/urban-development/audit/

[h]See http://ec.europa.eu/euroeurt/eurtistics-explained/index.php/Urban_Europe_%E2%80%94_eurtistics_on_cities,_towns_and_suburbs

[i]More information about the European Fund for Strategic Investments is available at: http://ec.europa.eu/growth/industry/innovation/funding/efsi_en

[j]Horizon 2020 is the financial instrument implementing the Innovation Union, a Europe 2020 flagship initiative aimed at securing Europe's global competitiveness. More information at: https://ec.europa.eu/programmes/horizon2020/

[k]The LIFE programme is the EU's funding instrument for the environment and climate action. The general objective of LIFE is to contribute to the implementation, updating and development of EU environmental and climate policy and legislation by co-financing projects with European added value

All information available at: http://ec.europa.eu/environment/life/about/

[l]Urban Innovative Actions (UIA) is an Initiative of the European Union that provides urban areas throughout Europe with resources to test new and unproven solutions to address urban challenges. Based on article 8 of the European Regional Development Fund (ERDF), the Initiative has a total ERDF budget of EUR 372 million for 2014–2020. Information available at: http://www.uia-initiative.eu/en

[m]See https://ec.europa.eu/eipp/desktop/en/index.html

[n]URBIS is a new dedicated urban investment advisory platform within the Hub. URBIS is set up to provide advisory support to urban authorities to facilitate, accelerate and unlock urban investment projects, programmes and platforms. More information available at: http://eiah.eib.org/about/initiative-urbis.htm

Table 8.2 European Union's urban initiatives on network, cooperation and awards

Networks and cooperation	Awards
• URBACT[a] • The Urban Development Network[b] • JPI Urban Europe[c] • European innovation partnership on smart cities and communities (EIP-SCC)[d] • Covenant of mayors for climate and energy[e] • International Urban Cooperation (IUC)[f]	• City Star (RegioStars Awards)[g] • European Capital of Innovation[h] • European Green Capital Award[i] • European Green Leaf Award[j] • Access City Award[k] • EU Horizon Prizes[l] • European Capital of Culture[m] • European Mobility Week Award[n] • The Europe 2020 Index: the progress of EU Countries, Regions and Cities to the 2020 targets[o]

Source Authors' own elaboration based on European Commission, https://ec.europa.eu/info/eu-regional-and-urban-development/topics/cities-and-urban-development_en

[a]URBACT Programme's mission is to enable cities to work together and develop integrated solutions to common urban challenges, by networking, learning from one another's experiences, drawing lessons and identifying good practices to improve urban policies. See http://urbact.eu/#

[b]The Urban Development Network is made up of more than 500 cities/urban areas across the EU responsible for implementing integrated actions based on Sustainable Urban Development strategies financed by ERDF in the 2014–2020 period. More information available at: http://ec.europa.eu/regional_policy/en/policy/themes/urban-development/network/

[c]See: https://jpi-urbaneurope.eu/

[d]The European innovation partnership on smart cities and communities (EIP-SCC) is an initiative supported by the European Commission that brings together cities, industry, small business (SMEs), banks, research units and others. It aims to improve urban life through more sustainable integrated solutions and addresses city-specific challenges from different policy areas such as energy, mobility and transport, and ICT. About the Marketplace of the European Innovation Partnership on Smart Cities and Communities, see https://eu-smartcities.eu/

[e]The Covenant of Mayors for Climate and Energy has brought together, since 2008, local and regional authorities who voluntarily commit to implementing the EU's climate and energy objectives in their territory. See https://www.covenantofmayors.eu/en/

[f]See http://ec.europa.eu/regional_policy/en/policy/cooperation/international/urban/

[g]European Union RegioStars Awards identify good practices in regional development and highlight original and innovative projects that are attractive and inspiring to other regions. See http://ec.europa.eu/regional_policy/EN/regio-eurrs-awards/

[h]See http://ec.europa.eu/research/innovation-union/index.cfm

[i]European Green Capital Award (EGCA), which recognises and rewards local efforts to improve the environment, the economy and the quality of life in cities. The EGCA is given each year to a city which is leading the way in environmentally friendly urban living and can thus act as a role model to inspire other cities. See http://ec.europa.eu/environment/europeangreencapital/index.html

[j]See http://ec.europa.eu/environment/europeangreencapital/europeangreenleaf/index.html

[k]The Access City Award recognises and celebrates a city's willingness, capability and efforts to ensure accessibility, see http://ec.europa.eu/social/main.jsp?catId=1141

[l]See https://ec.europa.eu/research/horizonprize/index.cfm

[m]The European Capitals of Culture initiative is designed to: (i) Highlight the richness and diversity of cultures in Europe; (ii) Celebrate the cultural features Europeans share; (iii) Increase European citizens' sense of belonging to a common cultural area; and (iv) Foster the contribution of culture to the development of cities: See https://ec.europa.eu/programmes/creative-europe/actions/capitals-culture_en

[n]The European Mobility Week Award has sought to improve public health and quality of life through promoting clean mobility and sustainable urban transport. See http://www.mobilityweek.eu/emw-awards/

[o]See http://ec.europa.eu/regional_policy/en/information/publications/regional-focus/2015/the-europe-2020-index-the-progress-of-eu-countries-regions-and-cities-to-the-2020-targets

8.4 The European Union's Urban Policy: Contributions to Territorial Cohesion and Their Transfer to the Everyday Life of European Cities and Regions

Territorial cohesion means closing the gap between different regions, and cities are both part of the problem and part of the solution to global challenges. On one hand, cities have a relevant role in regional growth, as the result of available human and capital resources (Glaeser 2000). Furthermore, they are also important for the development of rural areas. Indeed, according to Gagliardi and Percoco (2017), the positive effects of EU Cohesion Policy in rural areas can be boosted if these areas are close to a main urban agglomeration. However, due to their dimension, high-speed growth and high concentration of activities, they are faced with many social and environmental issues, which affect their citizens' quality of life (Table 8.3). The European Union's urban policy aims to solve problems and needs in urban areas and promote their growth potential (METIS 2016a).

During the period 2007–2013, activities undertaken under EU Cohesion Policy programmes related to urban development led to the achievement of several projects associated with urban regeneration to improve the business environment and the quality of life of European citizens in cities and regions. Table 8.4 reports the main outputs and outcomes attained and listed in the ex-post evaluation report of Urban Development and Social Infrastructures (METIS 2016b) and in the Seventh report on economic, social and territorial cohesion (European Commission 2017).

For the current programme period (2014–2020), EU urban investment priorities, under EU Cohesion Policy, are mainly focused on actions aiming to: (i) support a low-carbon economy; (ii) promote resource efficiency and climate change adaptation; and (iii) improve the urban environment and social inclusion (Table 8.5).

The aims of revitalisation of town centres and historic areas, construction or renovation of urban infrastructures, improve living ability in urban areas and make them more attractive for residents, tourists and investors—that characterised the EU 2007–2013 programming period—have been complemented for 2014–2020 with a set of new policy targets that increased significantly the EU's urban policy intervention rationale and have reinforced its capacity to, based on cities, contribute to a better achieving of the EU territorial cohesion objective.

Table 8.3 Cities: opportunities and barriers to economic and sustainable development

Opportunities	Barriers
• Firm concentration • Job opportunities • Higher education institutes	• Social challenges: poverty, social exclusion, lack of safety, insufficient infrastructure (e.g. transport and health care services) • Environmental issues: air quality, noise level and cleanliness

Source Authors' own elaboration based on Glaeser (2000) and EUKN (2011)

Table 8.4 Outputs and outcomes of EU urban development policy, 2007–2013

Outputs	Outcomes
• The construction, repair or renovation of infrastructure, such as schools, social and cultural centres, public spaces, health care services, streets and cycle paths	• Improved liveability in urban areas and in their neighbourhood
• Revitalisation of town centres, historic areas and river banks • Reuse of brownfield sites	• Making urban areas more attractive for residents and to attract more visitors and tourists
• Creation of business space	• Improving business environment and new job creation
• Installation of clean drinking water supply and wastewater treatment facilities	• Improving access to basic sanitation
• Improving buildings' energy efficiency	• Promoting Eco-efficiency

Source Authors' own elaboration based on METIS (2016b) and European Commission (2017)

Table 8.5 Objectives and related investment priorities of EU urban development policy 2014–2020

Thematic objective	Investment priorities
Low-carbon economy	• Promoting sustainable transport • Removing bottlenecks in key network infrastructures
Promotion of environmental protection, resource efficiency and climate change	• Promoting innovative technologies to improve environmental protection and resource efficiency (e.g. decrease the energy consumption of public buildings and additional capacity of renewable energy production) • Supporting industrial transition towards a resource-efficient economy, promoting green growth, eco-innovation and environmental performance management in the public and private sectors • Development of smart energy distribution
Better urban environment	• New and improved urban transport • Reduction of air pollution and promotion of noise-reduction measures • Revitalization of cities • Decontamination of brownfield sites, including conversion areas • Supporting business development and creation
Improvement of social inclusion and combating poverty	• Financing infrastructure for employment services • Investing in health and social infrastructure • Improving access to social, cultural and recreational services • Investing in education, training and vocational training for skills and lifelong learning

Source Authors' own elaboration based on European Commission (2011a), European Union (2013a, b) and European Parliament (2014)

The urban dimension of EU Cohesion Policy highlights the role of urban areas and cities in achieving the Europe 2020 objectives (European Parliament 2014), i.e. smart, sustainable and inclusive growth for the European Union.[11] For example, towards greener cities, three objectives of the Europe 2020 strategy are to: (i) reduce greenhouse gas emissions by at least 20% compared with 1990 levels; (ii) increase the share of renewable energy in final energy consumption to 20%; and (iii) move towards a 20% increase in energy efficiency (Nabielek et al. 2016). Furthermore, by the end of 2020, it is also expected that 41.2 million people will live in areas with integrated urban development strategies and 17,000 housing units will be redeveloped in urban areas (European Commission 2011a: 195), in order to improve city residents' quality of life. Indeed, according to a recent survey about the quality of life in European cities (European Union 2016), the two main dimensions with which European citizens are less satisfied are the availability of jobs and the affordability of housing.

8.5 The European Union's Urban Agenda in EU Cohesion Policy Post-2020

The EU's Urban Agenda was launched in May 2016 with the Pact of Amsterdam, with the aim to involve Urban Authorities in achieving 'Better Regulation', 'Better Funding' and 'Better Knowledge' (Table 8.6), and to contribute to the implementation of the United Nations 2030 Agenda for Sustainable Development, notably to make cities inclusive, safe, resilient and sustainable (European Commission 2016b). "The Urban Agenda for the EU will rely on the principle of an integrated approach to sustainable urban development as the guiding principle to achieve the goals of the three policy pillars" (European Commission 2016b: 6) described in Table 8.6.

Priorities, themes and cross-cutting issues for the EU's Urban Agenda (Table 8.7) were listed taking into account the priorities of the EU 2020 strategy for smart, sustainable and inclusive growth (European Commission 2016b: 7).

For the next long-term programming period 2021–2027 of EU Cohesion Policy, sustainable and integrated development of urban areas will remain as one of its priorities, as recently announced.[12] Sustainable, since several actions destined to support climate change and environmental objectives, such as promoting green infrastructure in the urban environment and reducing pollution, will still be an investment priority for urban areas, due to the greener, low-carbon Europe and climate adaptation policy objective. Integrated, because the instruments implemented for the 2014–2020 period, such as the Integrated Territorial Investment (ITI) and the Community-Led Local Development (CLLD) strategies, will be used to ensure the development of

---

[11] Smart means = Develop an economy based on knowledge and innovation; Sustainable in the sense of promote a more resource-efficient, greener and more competitive economy; Inclusive in order to foster a high-employment economy delivering social and territorial cohesion (European Commission 2010).

[12] COM (2018) 375 final and SWD(2018) 283 final.

Table 8.6 Achievement of EU's Urban Agenda

Better regulation	• More effective and coherent implementation of existing EU policies, legislation and instruments • It will not initiate new regulation, but will be regarded as an informal contribution to the design of future and revision of existing EU regulation
Better funding	• Identifying, supporting, integrating and improving traditional, innovative and user-friendly sources of funding for Urban Areas • It will not create new or increased EU funding, but will draw from and convey lessons learned on how to improve funding opportunities
Better knowledge	• Enhancing the knowledge base on urban issues and exchange of best practices and knowledge

Source Authors' own elaboration based on European Commission (2016a, b: 5–6)

Table 8.7 Priorities of EU's Urban Agenda and their contribution to smart, green and inclusive growth

Priorities of Urban Agenda	Smart	Green	Inclusive
Innovative and responsible public procurement	X		
Circular economy	X	X	
Urban mobility	X	X	X
Jobs and skills in the local economy	X		X
Digital transition	X		X
Air quality		X	
Energy transition		X	
Sustainable use of land and nature-based solutions		X	
Climate adaptation		X	
Affordable housing			X
Inclusion of migrants and refugees			X
Urban poverty			X

Source Authors' own elaboration based on Nabielek et al. (2016: 8)

urban areas, as highlighted in the Urban Agenda. New models of territorial governance, based on an integrated approach, could be a way to enhance the efficiency and effectiveness of financial instruments and to achieve the policy pillar of 'Better Funding'.

Future trends for the next programming period (2021–2027) also point to the need to reduce the administrative burden, namely, by simplifying the process of implementing EU funds (European Commission 2018a), which will lead to attaining the policy pillar of 'Better Regulation'. Additionally, ensuring data exchange about EU funds' beneficiaries and operations, through an open data platform (European Commission 2018a), will also make it possible to reach the 'Better Knowledge' pillar of Urban Agenda.

8.6 Final Remarks

The European Parliament considers that the smart specialisation model should become one of the leading approaches for the post-2020 Cohesion Policy phase, by encouraging cooperation between different regions, urban and rural areas and strengthening the EU's economic development (European Parliament 2017).

The relevance of smart specialisation strategies is "very strong as a public policy tool to support horizontal and vertical coordination between levels of administration and between policies. As well as for the strategic and functional articulation within the regional framework of ITI, CLLD and other territorial development approaches that may be defined for the new Cohesion Policy programming period, and to implement other territorial development strategies based on multi-policy approaches" (Neto et al. 2018: 21).

We propose the development of "a new stage for smart specialisation in the post-2020 Cohesion Policy, based on the increased social dimension and relevance of the RIS3. We also propose that this new generation of RIS3 should be designated RIS4—Research, Innovation and Social Strategies for Smart Specialisation" (Neto 2017: 22).

We consider desirable that in the future this new approach should be present in the definition of new urban policies, as it is more complete and prepared to fit the complexity of the problems and challenges facing cities. We also expect a new generation of urban public policies to be more effective in solving socio-economic and demographic problems, and generate an effect in other areas, namely, the environment, quality of life, mobility, urban pressure, urban regeneration and land use, among others, which are typical characteristics of urbanised societies.

The RIS4 approach we propose would also be very important for the operationalization of the new strategic priorities of the post-2020 Cohesion Policy such as globalisation, demography, migration, environment, climate change, security and defence, employment and digitalization of the economy and society, in the context of cities and urban policy.

The proposed RIS4 approach would provide a further opportunity to better plan the city as a complex system (which it is) and to ensure greater policy effectiveness, from a multisectoral and transpolicy perspective, with the policy instruments this new approach would bring.

Over the last 20 years, the EU has been able to build its *'Acquis Urbain'* and develop a European urban policy model based on a set of dimensions and areas of action with a high level of complementarity and systemic effects.

An European urban policy model based essentially on the following aspects: (i) creation and availability of financing instruments to support interventions in urban contexts; (ii) support for experimental, innovative and demonstrative urban projects; (iii) promotion of cooperation networks between cities; (iv) creation of systems for recognising and valuing good practices; (v) promotion of scientific research on topics relevant to urban issues; (vi) support for the creation, sharing and availability of information and knowledge to support decision-making in cities; (vii) progressive incorporation of urban issues into the different policies of the Union; and (viii) progressively incorporation of the EU's sectoral, technological, economic and societal strategic issues into urban policy.

The way in which the EU has come to understand and design its urban policy guidelines is even becoming increasingly broad, multisectoral and transpolicy, seeking to reconcile interventions in areas as diverse as infrastructure, mobility, culture, heritage, economy, employment, energy, ageing, technology, poverty and social inclusion, among many others. Therefore, the effort to promote territorial cohesion in the EU, together with the financial and other instruments that support it, should continue betting on the implementation of the EU Urban Agenda and, advantageously, be increasingly concentrated in cities, and in urban systems, in order to build and process, from them, the regional development of the territories in which they are located.

Acknowledgements The authors are grateful to Eduardo Medeiros, the editor, for the invitation to participate in this book.

References

Atkinson R (2014) The urban dimension in cohesion policy: past developments and future prospects. In: Paper presented at an RSA workshop on 'The New Cycle of the Cohesion Policy in 2014-2020'. Institute for European Studies, Vrije Universiteit Brussels, 24 March 2014

Barca F (2009) An agenda for a reformed cohesion policy. A place-based approach to meeting European Union challenges and expectations. Independent report prepared at the request of Danuta Hübner, Commissioner for Regional Policy

De Bruijn M, Zuber P, Piskorz W, Rózsa J (2015) Scenarios for integrated territorial investments. In: European Commission, Directorate-General for Regional Policy. Publications Office of the European Union, Luxembourg

Dijkstra L, Athanasoglou S (2015) The Europe 2020 index: the progress of EU countries, regions and cities to the 2020 targets. European Commission, Regional and Urban Policy, Regional Focus 01/2015, May 2015

EUKN (2011) Cohesion policy in urban practice. European Urban Knowledge Network, The Netherlands

European Commission (1997) Towards an urban agenda in the European Union, COM (97) 197 final

European Commission (1998) Sustainable urban development in the European Union: a framework for action, COM (1998) 605 final

European Commission (2006) Cohesion policy and cities: the urban contribution to growth and jobs in the regions. COM (2006) 385 final

European Commission (2007) The urban dimension in community policies for the period 2007-2013. European Commission, Directorate General Regional Policy, Brussels

European Commission (2008) Fostering the urban dimension: analysis of the operational programmes co-financed by the European Regional Development Fund (2007-2013). Working Document of the Directorate General for Regional Policy, Brussels, 25 November 2008

European Commission (2009) Promoting sustainable urban development in Europe. Achievements and opportunities. European Commission, Directorate-General for Regional Policy. Publications Office of the European Union, Luxembourg. https://doi.org/10.2776/85168

European Commission (2010) Europe 2020—a European strategy for smart, sustainable and inclusive growth. COM (2010) 2020 final
European Commission (2011a) Cohesion policy 2014-2020, investing in growth and jobs. European Commission, Directorate-General for Regional Policy, Belgium
European Commission (2011b) Cities of tomorrow—challenges, visions, ways forward. European Commission, Directorate-General for Regional Policy. Publications Office of the European Union, Luxembourg
European Commission (2014) The urban dimension of EU policies—key features of an EU urban agenda. COM (2014) 490 final
European Commission (2016a) The State of European Cities 2016. Cities leading the way to a better future. European Commission, Directorate-General for Regional and Urban Policy. Publications Office of the European Union, Luxembourg. https://doi.org/10.2776/643506
European Commission (2016b) Establishing the urban agenda for the EU 'Pact of Amsterdam'. Agreed at the Informal Meeting of EU Ministers Responsible for Urban Matters on 30 May 2016 in Amsterdam
European Commission (2017) Seventh report on economic, social and territorial cohesion. European Commission, Directorate-General for Regional and Urban Policy, Belgium. https://doi.org/10.2776/176864
European Commission (2018a) SWD(2018) 283 final—staff working document executive summary of the impact assessment accompanying the document proposals for a regulation of the European Parliament and of the Council on the European Regional Development Fund and on the Cohesion Fund on a mechanism to resolve legal and administrative obstacles in a cross-border context on specific provisions for the European territorial cooperation goal (Interreg) supported by the European Regional Development Fund and external financing instruments, Strasbourg
European Commission (2018b) Proposal for a regulation of the European Parliament and of the Council laying down common provisions on the European Regional Development Fund, the European Social Fund Plus, the Cohesion Fund, and the European Maritime and Fisheries Fund and financial rules for those and for the Asylum and Migration Fund, the Internal Security Fund and the Border Management and Visa Instrument. COM (2018) 375 final
European Commission (2018c) Staff working document executive summary of the impact assessment accompanying the document Proposals for a Regulation of the European Parliament and of the Council on the European Regional Development Fund and on the Cohesion Fund on a mechanism to resolve legal and administrative obstacles in a cross-border context on specific provisions for the European territorial cooperation goal (Interreg) supported by the European Regional Development Fund and external financing instruments. SWD(2018) 283 final
European Parliament (2014) The role of cities in cohesion policy 2014-2020. European Parliament's Committee on Regional Development, Belgium
European Parliament (2017) Report on building blocks for a post-2020 EU cohesion policy (2016/2326(INI)), Committee on Regional Development
European Union (2013a) Regulation (EU) N° 1300/2013 of the European Parliament and of the Council of 17 December 2013 on the Cohesion Fund and repealing Council Regulation (EC) No 1084/2006. Off J Eur Union L 347:281–288 (20.12.2013)
European Union (2013b) Regulation (EU) N° 1301/2013 of European Parliament and of the Council of 17 December 2013 on the European Regional Development Fund and on specific provisions concerning the Investment for growth and jobs goal and repealing Regulation (EC) No 1080/2006. Off J Eur Union L 347:289–302 (20.12.2013)
European Union (2016) Quality of life in European cities 2015. Publications Office of the European Union, Luxembourg
Ferry M, McMaster I, Zwet A (2018) Assessing the performance of integrated territorial and urban strategies: challenges, emerging approaches and options for the future. European Commission, Directorate-General for Regional and Urban Policy and European Policies Research Centre (EPRC). Publications Office of the European Union, Luxembourg. https://doi.org/10.2776/96617

Gagliardi L, Percoco M (2017) The impact of European cohesion policy in urban and rural regions. Reg Stud 51(6):857–868. https://doi.org/10.1080/00343404.2016.1179384

Glaeser EL (2000) The new economics of urban and regional growth. In: Clark GL, Feldman MP, Gertler MS, Williams K (eds) The Oxford handbook of economic geography. Oxford University Press, pp 83–98

Lavalle C, Pontarollo N, Batista e Silva F, Baranzelli C, Jacobs C, Kavalov B, Kompil M, Perpiña Castillo C, Vizcaino M, Ribeiro Barranco R, Vandecasteele I, Pinto Nunes Nogueira Diogo V, Aurambout J, Serpieri C, Marín Herrera M, Rosina K, Ronchi S, Auteri D (2017) European territorial trends—facts and prospects for cities and regions. European Commission, Joint Research Centre. Publications Office of the European Union, Luxembourg. https://doi.org/10.2760/148283

Medeiros E (2016) Is there a rise of the territorial dimension in the EU Cohesion Policy? Finisterra, LI 103(2016):89–112. https://doi.org/10.18055/Finis7940

Medina MG, Fedeli V (2015) Exploring European urban policy: towards an EU-national urban agenda? Gestión y Análisis de Políticas Públicas, Nueva Época, no 14 julio-diciembre 2015

METIS (2016a) The growth potential of an integrated EU urban agenda. European Committee of the Regions, Belgium

METIS (2016b) Ex post evaluation of urban development and social infrastructures—final report. Work Package 10 in Ex post evaluation of Cohesion Policy programmes 2007-2013, focusing on the European Regional Development Fund (ERDF) and the Cohesion Fund (CF), European Commission, Directorate-General for Regional and Urban Policy, Belgium

Nabielek K, Hamers D, Evers D (2016) Cities in Europe—facts and figures on cities and urban areas, PBL Netherlands Environmental Assessment Agency. PBL Publishers, The Hague

Neto P (2014) Identity and plasticity in the reshaping processes of urban landscape. In: Proceedings of the 10th meeting of the workshops of the council of Europe for the implementation of the European landscape convention, Évora, Portugal, 20-21 octobre 2011. European spatial planning and landscape, no. 97. Council of Europe Publishing, Strasbourg, pp 81–91

Neto P (2017) Post-2020 cohesion policy and the new challenges for public policy, conference presented, as keynote speaker, at the official dinner of the international seminar on social services of general interest and territorial cohesion: experiences and challenges, organized by the Portuguese Agency for Development and Cohesion (AD&C) I.P., Évora Hotel, November 13, 2017, Évora, Portugal

Neto P, Serrano MM, Santos A (2018) Renewed challenges for public policies in post-2020 cohesion policy. From RIS3 to RIS4 and a new social dimension for smart specialisation. Publ Policy Portuguese J 3(1):7–25

Zwet A, Bachtler J, Ferry M, McMaster I, Miller S (2017) Integrated territorial and urban strategies: how are ESIF adding value in 2014–2020? European Commission, Directorate-General for Regional and Urban Policy and European Policies Research Centre (EPRC). Publications Office of the European Union, Luxembourg. https://doi.org/10.2776/50425

Paulo Neto is a Professor in the Department of Economics at the University of Évora, Portugal. He has a Ph.D. and Aggregation in Economics and is the author of several books and articles published in both national and international scientific publications in the field of Regional and Urban Economics, Public Policies and European Economics. He is Coordinator of the UMP-P—Public Policy Monitoring Unit at the University of Évora, Director of the Public Policy Portuguese Journal, Coordinator of the Portuguese Network of Research in Public Policies, member of the Representative Assembly of the Order of Economists in Portugal and a Researcher at CICS.NOVA.UÉvora.

Maria Manuel Serrano has a Ph.D. in the Sociology of Economics and Organisations and a master's in Socio-organisational Systems of Economic Activity from ISEG/UL—Institute of Economics and Management/University of Lisbon and a degree in Sociology from the University of Évora. She is a Professor at the University of Évora and Course Director of the Degree in Sociology. She is a Researcher at CICS.NOVA.UÉvora—Interdisciplinary Centre for Social Sciences and a Collaborating Researcher at SOCIUS/CSG—Centre for Research in the Sociology of Economics and Organisations. She is Sector Coordinator of UMPP—Public Policy Monitoring Unit at the University of Évora and Director of the Public Policy Portuguese Journal.

Anabela Santos is a Research Fellow at the International Centre for Innovation, Technology and Education Studies (iCite) of the Solvay Brussels School of Economics and Management at Université Libre de Bruxelles (ULB) in Belgium. From 2014, she has been a team member of UMPP—Public Policy Monitoring Unit at the University of Évora in Portugal. She has been involved in several research studies about the effectiveness of public support in the regional context in Portugal. She was also financial advisor for small- and medium-sized enterprises located in a rural area in the Portuguese Alentejo region for a period of 9 years.

Chapter 9
Towards a European Geography of Future Perspectives: A Story of Urban Concentration

Kai Böhme, Christian Lüer and Maria Toptsidou

Abstract Urban areas and the diversity of urban areas across Europe are a corner stone of territorial cohesion. At the same time, territorial cohesion in Europe is challenged by the dominant role of urban areas in territorial development and in territorial development policies. One may conclude that the urban perspective and territorial cohesion cannot do without each other, but at the same time, they risk to wreck each other. This contribution will further explore this complicated relation. Starting with the growing importance of urban areas, the first section will address the increasing concentration of economic and demographic growth in Europe to urban areas (Böhme and Lüer 2017; ESPON 2017, 2018). Based on the idea that cities act as motors for territorial development in a wider region, over the past decades they have become the centre points for development and future perspectives. The impact this had on territorial cohesion can be discussed at least at two different levels: (a) European and (b) functional urban/regional level.

Keywords Urban development · Urban diversity · Functional urban areas Territorial development · Territorial cohesion · Fragmentation · Europe Concentration · Future · Foresight · Future perspectives

9.1 Introduction

Urban areas have always played an important role in the societal, economic and territorial development in Europe. Compared to other continents, Europe is characterised by a rather polycentric and less polarised urban fabric, and the diversity of

K. Böhme (✉) · C. Lüer · M. Toptsidou
Spatial Foresight, Heisdorf, Luxembourg
e-mail: kai.boehme@spatialforesight.eu

C. Lüer
e-mail: christian.lueer@spatialforesight.eu

M. Toptsidou
e-mail: maria.toptsidou@spatialforesight.eu

E. Medeiros (ed.), *Territorial Cohesion*, The Urban Book Series,
https://doi.org/10.1007/978-3-030-03386-6_9

urban areas of different sizes across Europe is a corner stone of territorial cohesion, as outlined by various ESPON studies (e.g. 2014a).

The importance of urban areas for Europe is also well echoed in European policy making. Several European policy documents have emphasised the importance of cities, such as the Territorial Agenda of the European Union 2020 (MSPTD 2011), the Leipzig Charter (European Commission 2007) and the EU Urban Agenda (MUMEU 2016), which aims at improving horizontal coordination between EU sector policies and vertical coordination among European, national and local governments. Also, the urban earmarking in EU Cohesion Policy is a clear indication of the importance given to urban areas in policy making.

At the same time, territorial cohesion—and even integration—in Europe is challenged by the dominant role of urban areas in territorial development and in territorial development policies. One may conclude that the urban perspective and territorial cohesion cannot do without each other, but at the same time, they risk to wreck each other. This contribution will further explore this complicated relation.

To elaborate on this point, we first provide some more background on the role urban areas play in Europe and on their diversity in terms of profiles and characteristics. The point made is that urban areas are centre points of development and a lot of our activities are concentrated on them.

Looking towards the future, we will explore a number of development trends, which are likely to influence societal, economic and territorial development in Europe in the decades to come. Although the discussion of possible future developments is linked to huge uncertainties, a distinct territorial feature seems almost to be certain: Europe's future will be urban and bring about an even stronger concentration to fewer urban areas as hotspots.

This increasing concentration to urban areas is an expression of a societal, economic and territorial fragmentation, and further accelerates this fragmentation and territorial inequalities. This leads to rather different development outlooks for different parts of European society and territory. We argue in this contribution that it actually leads to a territorially rather diverse 'European geography of future perspectives'.

Closing off, we explore different alternative futures for Europe which might be able to bridge the increasing fragmentation. The reader may find that some of them are rather wild and unlikely, while others sound like only minor adjustments though they still might be tough to implement. In any case, if we do not want the increasing concentration on urban areas to undermine the fundament of economic development and integration in Europe, we may need to pay more attention to societal and territorial fragmentation trends and implications our policy choices and scenarios may have on regions, cities and the overall territorial balance in Europe (Böhme and Lüer 2017; Böhme and Toptsidou 2017).

9.2 Urban Patterns and Diversities in Europe

Urban areas are a corner stone of economic development and territorial cohesion in Europe. The urban pattern of Europe and the diversity of urban areas across Europe play a vital role in this. Furthermore, cities are widely perceived as motors for development leading to increasing economic and demographic concentration towards urban areas in Europe. In the following section, we will discuss Europe's territorial pattern and diversity of urban areas.

9.2.1 Europe's Urban Pattern

There are different ways of defining and counting cities and urban areas. Following the approach used by ESPON, the territory of the EU plus Iceland, Liechtenstein, Switzerland and Norway is home to almost 700 cities with 50,000–250,000 habitants and about 100 metropoles with more than 250,000 inhabitants. 72% of the EU28 population live in urban areas, with 41% living in cities of more than 50,000 inhabitants and another 31% in intermediate urbanised towns and suburbs. The urban areas altogether, however, account for only 17% of the territory (PBL Netherlands Environmental Assessment Agency 2016: 12). The share of the population living in these urban areas is expected to further increase and projected to reach just over 80% by 2050 (Eurostat 2016: 9).

Focusing on the largest areas, metropolitan regions with more than 250,000 inhabitants (including their commuting zones) are home to 59% of the population of the EU. The demographic weight of these areas is even topped when looking at economic figures 62% of the jobs in Europe are located and 68% of the GDP is produced in such functional urban areas (European Commission and UN-Habitat 2016: 58). Capital areas play a particular role in this as they often are home to a more highly educated population, higher innovation scores and higher productivity levels (ibid.).

This idea of urban areas as motors, engines or hotspots of development has been nourished both in research and policy making in the past decade. Nevertheless, the narrative of concentration to few hotspots as motors for future development, as promulgated by urban economics, does not always hold because urban scale effects are not ubiquitous (Dijkstra et al. 2013). Not all metropolitan regions are frontrunners and 'motors of development' in their respective countries. Growth rates of various metropolitan areas are only similar to, or even considerably below, those of their countries (OECD 2006).

The overall urban dominance is shared among a vast number of cities and larger metropolitan areas. Compared to other continents, Europe is characterised by a rather polycentric and less polarised urban fabric (ESPON 2014a), where important urban centres are geographically distributed and spread across Europe. This is among others illustrated by the large number of major cities, which are located in comparatively close proximity to each other (compared to other continents), yet the closest large city is further away than in other parts of the world (European Commission and UN-

Habitat 2016). This is also the outcome of a dense network of small and medium-sized urban areas with a population of 5,000–50,000 inhabitants, which play an important role in the urban fabric with regard to balanced development—a much greater role than on other continents (Barca et al. 2012). About 8,350 of them exist in the EU28 being home to almost 25% of the population (Latvian Presidency of the Council of the European Union 2015: 6).

9.2.2 Europe's Urban Diversity

Even more than on the territorial pattern, Europe's polycentricity depends on the diversity of cities, their profiles, specialisations and the functions they cover for their surroundings and wider territories. Each urban area in Europe has a different functional profile, defining their role in the European urban system. ESPON developed an analysis of this functional diversity composing of six different types of functions (2005, 2007):

(a) **Administrative functions** in the public administrative system at international, national or regional level: The current European pattern mainly reflects the hierarchies within individual national systems with the national capitals as main nodes. Non-capital cities hosting institutions that are part of the EU institutional system (e.g. Strasbourg, Frankfurt) or other international organisations complement the picture.

(b) **Decision functions** due to headquarters and subsidiaries of important private companies: Many headquarters in Europe are located within the European core, the so-called 'Pentagon', an area comprising of southeast England (including London), the Benelux countries, western Germany, northeast France (including Paris) and northern Italy.

(c) **Transport functions** to connect urban areas: In many small countries, only one transport-oriented city (i.e. a focus on international aviation and container harbours) exist. The most important transport nodes of European and global relevance are again located within the 'Pentagon'.

(d) **Knowledge functions** due to universities and research centres: Research centres and universities are historically distributed across different parts both at European and national level. Also, smaller urban centres play an important role. Hence, the European picture is rather balanced.

(e) **Tourism functions** due to natural and cultural assets: Besides the capitals and other major European cities increasingly visited by city and weekend tourists, especially urban areas in the Alps and the Mediterranean area have a clear focus on touristic functions.

(f) **Industrial and manufacturing functions** as economic backbones: The national urban systems are often the result of industrialisation. Urban areas with a strong focus on this function are within the 'Pentagon'. However, many old-industrialised regions face challenges as traditional industries are affected by the structural change and are in decline.

The analysis shows that different metropolitan areas have different functional specialisations, i.e. they strengthen specific functions and sharpen or adjust their individual profiles (ESPON 2005). This specialisation helps municipalities become more visible and play a stronger role in the fabric of urban areas, which increases the diversity of the urban pattern. Differences in specialisation can, therefore, be considered a driving force of the European polycentric urban system.

Regardless of their role or specialisation in the European urban system, larger cities benefit from agglomeration advantages. They act as gateways for all kinds of flows and functions, from business and transport to knowledge and attraction, within larger networks at global, EU, national and regional scale (ESPON 2013).

In contrast to this, smaller urban areas depend on a clear profile and the exploitation of comparative advantages. One way small urban areas realise these advantages is cooperation, i.e. they develop and maintain good relations with other urban and rural areas. They hence depend on their capacity and willingness to engage in collaborative and cooperative actions with other areas, e.g. to develop joint projects or share services (ESPON 2014b).

The functional profile of an urban area is not static. It changes over time and can be influenced by policy choices. Nowadays, many urban areas are laboratories for new policy approaches and are increasingly characterised by a focus on experimentation, testing and 'learning by doing' in all kinds of policy fields, from technological innovation and 'smart cities' to social innovation and sustainability. Trends like decentralisation, increasing autonomy for local authorities and more involvement of citizens in policy formulation and implementation have supported this development (European Commission and UN-Habitat 2016). The flexible profile of urban areas allows for developing new and amending existing functions and profiles, and will consequently further promote Europe's urban diversity.

9.3 As for the Future

The above-sketched territorial imbalance, i.e. dominance of urban areas, is likely to stay for a foreseeable future. The territorial pattern of Europe is unlikely to change over the next decades, i.e. urban areas stay where they are. However, what we do in places, the profile and functions of places, the links and interactions between places may change rather rapidly. These places, links and interactions are influenced by a wide range of different development trends, which, in turn, will also influence the future territorial balance in Europe, be it within urban areas, between larger and smaller urban areas, or between non-urban areas and urban areas.

There are various different attempts to collect and categorise trends (Böhme et al. 2016; Böhme and Lüer 2016, 2017; ESPON 2018; European Strategy and Policy Analysis System 2015). One distinction identifies exogenous and endogenous trends.

While the former have to be taken into account as general framework conditions and are most difficult for policy and decision makers to influence directly, endogenous trends are subject to, or even direct expression of, policy choices and political priorities.

This distinction can be combined with different thematic fields which help to further distinguish the trends. Technological and environmental trends are mainly exogenous trends, which refer to long-term, often global developments with enormous influence on the other fields. The Fourth Industrial Revolution and climate change are two prominent examples of trends that are expected to have rather disruptive and comprehensive impacts which might, however, differ between territories. Some other trends in the societal and economic spheres are of more endogenous character, yet have a certain degree of autonomy, e.g. internal and external migration flows, polarisation between autonomous and open regions or the circular economy model. Finally, political trends are mainly endogenous trends and are matter to the course and outcome of democratic decision-making processes.

Another important characteristic refers to the level where the trends occur. Here, an additional distinction can be made between trends that are more 'domestic' European trends and other trends, which are of global character or at least underline that what happens elsewhere on the globe has impacts on territorial development in Europe.

The figure below (Fig. 9.1) combines these three distinctions (thematic fields, exogenous versus endogenous, global versus European) and provides a non-exhaustive overview of a variety of trends that are relevant for future territorial development in Europe. More exogenous thematic fields are at the top of the figure, more endogenous thematic fields at the bottom. Thematic fields can be distinguished by colours. Global trends are displayed in darker colour and European trends in lighter colour. Of all these trends, two aspects might be particularly vital for the further concentration of developments in urban areas. First, the technological and related economic trends in the sphere of the Fourth Industrial Revolution, including also the shift to data economy, platform economy and monopolisation. Second, demographic trends including ageing and migration.

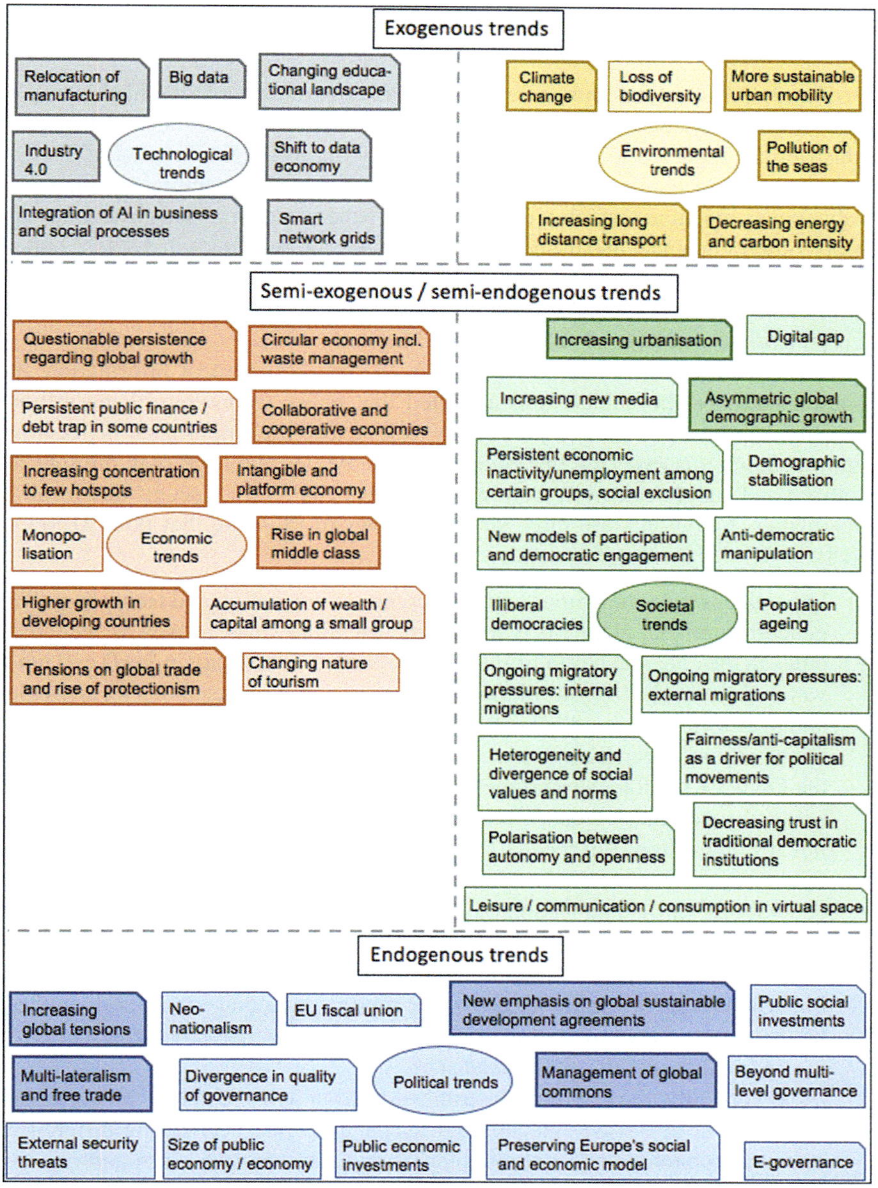

Fig. 9.1 Overview of global and European exogenous and endogenous trends. *Source* Authors

9.3.1 More Urban Focus Due to Technological Progress

The Fourth Industrial Revolution describes the fusions of technologies blurring of the lines between physical, digital and biological systems. In the next decade, firms and industries are expected to become predominantly digitised. The Internet of Things, supported by big data analytics, artificial intelligence and machine learning will enable smart machines which are increasingly benefitting from sensor technology, cheap computing power and the real-time use of algorithms. This will disclose opportunities for new business models and entrants, together with new challenges concerning the substitution of labour and the role of the European economy in new patterns of production at world level (ESPON 2018; Schwab 2017).

As in earlier Industrial Revolutions, e.g. the breakthrough of information and communication technologies, it is expected that major technological innovations including robotics and fusion technology will reduce the importance of location. Among others, it is expected that production can be decentralised (e.g. through 3D printing) and people can work from wherever they like. Still, it seems that location does matter. Below just are three short examples to illustrate that point (ESPON 2018):

(1) Urban areas hold the primacy when it comes to innovation and technology. The general European picture is that large metropolitan areas, as well as secondary cities in the core and North of Europe, are particularly strong in R&D and innovation (Lüer et al. 2015). Looking more particularly at key enabling technologies and SMEs, the geographical pattern across Europe is somewhat more balanced. Especially in larger European countries like France, Germany, Italy, Spain and the United Kingdom, several metropolitan regions have an innovative capacity below the national average, hinting at the complexity of spatial structures in the field of innovation (European Commission and UN-Habitat 2016). A particular field refers to green tech innovations, which are even more concentrated than general innovation and technology locations: Metropolitan regions and capital regions mainly in Western Europe are hotspots for green tech (ESPON 2018). As green growth, green economy and circular economy are expected to be an important pillar of our future economic model, particular innovations in these fields may hold the key for future developments. This may give areas currently leading in these fields a clear head start.

(2) More generally, the Fourth Industrial Revolution will accelerate territorial differences. The next wave of technological changes will exacerbate the differences between technological/economic players and also between cities and regions. Following current developments in the technology sectors and platform economy, also for the next Industrial Revolution the advantages will be on early adapters as 'the winner takes it all'. Best performers are expected to capture a very large share of the rewards, and the remaining competitors are left with very little (Réchard et al. 2016). These winners certainly have locations and these may very well matter, as some places are able to adjust faster to the new technological advancements and their needs. This is the case in particular when

it comes to different technology environments, legal frameworks, social acceptance and economic pressure. According to some experts (e.g. Schwab 2017), the five cities with the most effective policy environment to foster innovation are New York, London, Helsinki, Barcelona and Amsterdam. So, Europe might host four of the top five locations for developments to come. Estonia is an example of a country that follows the vision of a digital republic ('e-Estonia'), applying a broad participatory approach involving the public and the private sector as well as citizens. As a forerunner in the digital sector, it has brought about various start-ups over the past years and also attracts companies from other places to relocate to Estonia.

(3) Technology may decrease agglomeration advantages and increase sprawl. Urban sprawl is expected to increase as new forms of mobility paired with decentralisation possibilities reduce agglomeration advantages. (ESPON 2018) Advances in service robotics, 3D printing and logistics technology are already reducing costs in manufacturing and service, allowing small-scale production to be profitable. The expected levels of automation of many tasks (e.g. by service robots) could lead to a sharp reduction in the minimum efficient scale for many businesses, especially those providing consumer-oriented services. However, urban sprawl will not be linear. Instead, the growth of households in rural and exurban areas compared to cities is likely to start to look like a barbell. On one end of the barbell, some cities will continue to grow successfully, attracting the wealthy, the young and empty nesters who choose urban cores as their places of living without financial constraints. On the other end are residential zones at the edge of metropolitan regions that offer living space at lower costs and attract middle-class families who will blend the features of rural living with urban amenities more easily due to declining costs of distance. Traditional suburbs and small and medium-sized towns might be trapped in the middle as spaces in-between without real appeal facing population decrease. This might become the basic territorial structure for a 'post-urban economy', in which commuting distances no longer matter and people rather choose where to live based on lifestyle characteristics and amenities (e.g. good weather, vibrant social and cultural offers, proximity to recreational activities, family or peers) (Bain and Company 2016).

Summing up, trends in technological change are expected to substantially increase concentration of the locational advantages and 'places to be' to a very few hotspots following the philosophy of 'the winner takes it all'. So, one territorial face of the next Industrial Revolution concentration will be concentration at a global and European level to a much higher degree than what we have seen in the past decades. At the same time, at local level, location may matter less than today and we may see more decentralisation probably mainly at the edge of thriving urban and metropolitan areas. So, urban sprawl will be the other territorial face of the next industrial revolution at local and regional level.

9.3.2 More Urban Focus Due to Demographic Change

Europe faces an increasing influence of demographic change. Territorial concentration of the population in different areas in Europe seems then inevitable, as ageing population together with migration patterns point at a concentration towards urban areas, which leave some regions with severe demographic challenges (Böhme and Lüer 2017; ESPON 2017, 2018). Different aspects and dimensions will influence this concentration and shape demographic change over the next decades, five of which are shortly introduced in the following:

(1) One dimension of this concentration is the polarisation between metropolitan and non-metropolitan regions (see text box), expected to increase over the next years and resulting in a stronger urban–rural divide. Furthermore, there are trends favouring the population growth in urban centres especially in Western and Northern Europe (ESPON 2018). This is especially to be seen in large metropolitan areas in northwest Europe and in the Baltic Sea Region. At the same time, this poses challenges in non-urban parts of Eastern European countries and Germany, which continue to decline resulting in the phenomenon of shrinking regions (ESPON 2017).

(2) The concentration of population growth in urban centres is accompanied by the effects of a territorial imbalanced age structure, with urban centres being characterised by ageing population. Europe is an ageing continent and will continue to be so in future. The combination of an improved quality of life with low fertility rates poses Europe at risk of a silver tsunami. Such a trend poses social and economic challenges for some regions, such as healthcare demands, decreasing labour force and pension systems (ESPON 2017).

(3) Intra-regional and intra-European migration is another aspect to take into account that accelerates the demographic concentration in urban areas. Intra-regional and national migration regards migration of people to urban centres within the same region or country, while intra-European migration is about people moving from one country to another within Europe. Intra-regional migration reflects the 'rural exodus', where people move from remote rural centres to more attractive urban centres. Hence, dynamic urban centres report high positive net migration, with capital cities showing the highest rates. Similarly, intra-European migration is also more visible in urban centres, where the young and talented move for higher education and better job prospects (ESPON 2017).

(4) Europe is also closely tied to global migration flows and these flows are expected to grow in the years to come. Inequalities in terms of population and wealth, as well as armed conflicts and climate change but also simply age structures will accelerate global migration in various directions. Increased migration could mean up to 400 million people being on the move by 2050, including more climate change refugees (Glenn et al. 2015). While Europe is ageing and faces challenges of adjusting to increasing median ages, young and ambitious societies with a low median age and very young populations face different challenges. More young people need to find a place in society in competition with both those

who are settled in positions of power and with other young people also striving to advance their careers. Better education and limited resources are additional ingredients in this mix, which could easily lead to multiple conflicts. While the share of young people declines in Europe, it grows in Africa and is stable in Asia. This may very well push more people to search their way to Europe, and there mainly to the urban areas (National Intelligence Council 2012).

(5) While Europe attracts migrants from its neighbourhood, resulting in continuous migrant flows across the Mediterranean, an increasing number of young and talented Europeans will consider vibrant urban centres outside Europe as more attractive places to work and live than ageing European cities. Europe will face increasing difficulties to stop a brain drain to fast-growing economic, social and cultural centres outside Europe—and new centres may emerge given the growing number of highly skilled young people outside Europe. In the global competition for experts, these centres will increasingly become competitors to economic urban agglomerations in Europe (Böhme et al. 2016; Lüer et al. 2015).

Summing up, the overall trends of an ageing population and migration patterns in favour of urban areas, leave some regions with severe demographic problems, and are expected to increase rural–urban disparities in Europe.

Text Box—Change of Demographic Potential Over Time

Population potential is the total number of persons living within a certain distance of a given place. Analysing population data at the level of 7 million 1 km^2 grid cells rather than administrative units, Spatial Foresight could establish information on demographic change showing changes in the number of people living within 50 km of each cell between 1961 and 2011. For the calculation, 50 km as a proxy for a daily commuting distance was used. Cutting a long story short, the analysis shows at a very detailed level various phases of urban–rural polarisation (Fig. 9.2). In particular, in the early 2000s, the population potential decreases steadily in some remote and rural areas, e.g. along the axis running from Lorraine to Auvergne in France or in remote parts of northern Scandinavia. It seems that efforts to mitigate depopulation in these areas can, at best, postpone trends that may appear as inevitable and alleviate their social implications.

9.4 Concentration on Urban Areas: Necessity or Challenge?

As shown above, the current territorial patterns of Europe are characterised by the strong dominance of urban areas. This dominance is most likely to become even

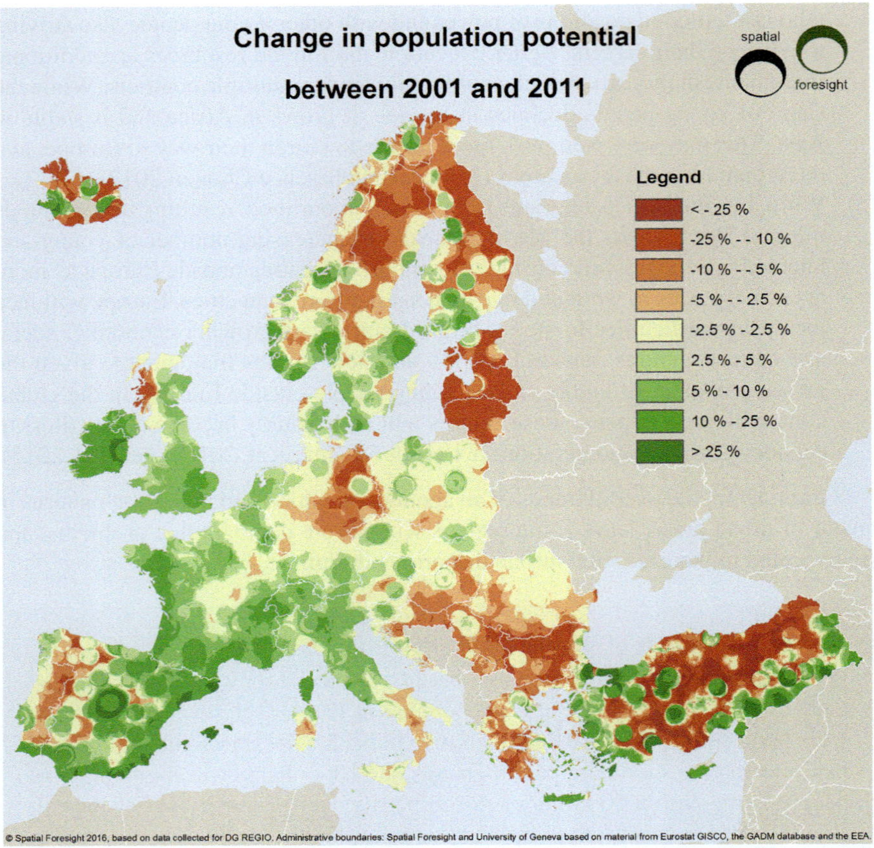

Fig. 9.2 Change in population potential in Europe between 2001 and 2011. *Source* Spatial Foresight

stronger in the years to come, as major development trends suggest an increasing concentration to (a selected number of) urban areas. This makes perfect sense in economic terms and economics might even suggest that Europe needs to further strengthen the concentration on urban areas. Urban areas stick out in terms of innovation, productivity, advancements in the service sector and global competitiveness. So, for Europe to prosper and have a strong stand in the global economy, the concentration on urban areas might almost be considered as imperative. This is very much in line with the idea of 'cities as engines for development' referred to earlier.

The flipside of the coin portrays the situation of non-metropolitan areas in Europe. Territorial cohesion and polycentric development might not be far, however, the idea of territorial cohesion is becoming more distant (ESPON 2018). This impact on territorial cohesion can be discussed at least at two different levels: (a) European and (b) functional urban/regional level, as urban policies go beyond administrative borders.

At European level one may argue that concentration to urban areas may have contributed to a sort of territorial balance of urban areas throughout Europe. Even in more sparsely populated areas, the focus on urban areas implied that there is at least an urban area which is notable at European scale and thereby contributes to the territorial balance across Europe.

On the other hand, an increasing concentration to urban areas also implies increasing territorial disparities in Europe. One may argue that concentration goes beyond territorial disparities and successively translates into societal fragmentation. Concentration processes will further spur Europe's changing economic geography including increasing fragmentation of society and territories. While this fragmentation is a consequence of economic and social developments in Europe, it also puts Europe's potential to respond to globalisation at risk.

At functional urban/regional level one may argue that economic success was always centred around certain places and not spread evenly and it would, therefore, be more sensible to focus on thriving areas, expand them and rather support internal migration to these places. For the British economic powerhouse of London, for example Leunig suggested to make London a mile bigger and this way create space for 400,000 new houses (Leunig 2008).

On the other hand, metropolitan regions already today look more like metropolitan regions from other countries than other parts of their own countries. If wealth is increasingly concentrated to specific groups in society and fewer hotspots, this will also lead to increasingly stronger intra-regional inequalities. These inequalities influence opportunities of residents in lagging areas and lead to the underutilisation of regional potentials (World Bank 2018). Eventually, the societal, economic and territorial fragmentation and the political consequences thereof, may undermine the fundament of a functional urban/regional area as a whole.

As afore mentioned, concentration processes are expected to be, i.e. the consequence of the philosophy of 'the winner takes it all' implied by different trends. A European Single Market in which only few countries or regions take all the benefits of the (economic) integration process, while others suffer from the adverse consequences, would, however, widely ignore the objectives of both social and territorial cohesion. Such an outcome of European integration would, therefore, not be consistent with the most fundamental philosophical values of the EU (Barca et al. 2012).

Key EU policies—beyond the realm of Cohesion Policy—have supported efforts to continuously develop political, social and cultural cohesion, and to reduce barriers to achieving this. Examples are the European Single Market and trans-European transport networks (TEN-T). Still, Europe faces many different types and complexities of increasing fragmentation, with some as cause and others as consequence of a range of trends. These complexities include political choices, the social fabric within and across different regional contexts, geographical landscapes and economic circumstances in different locations. Inequality between and within countries has been recognised as a crucial factor driving fragmentation, including political instability and the rise of populist movements and parties—referred to as the revenge of places that do not matter (Rodríguez-Pose 2017).

Consequently, different places face different 'everyday realities' and different future prospects, implying a strongly territorially differentiated 'geography of future perspectives'. So far, we got used to that different social groups see different future potential or challenges and some groups see actually very little hope for a future which allows them to get out of misery. Today, we see increasingly that the same holds true for different types of territories or places. Whereas some places—meaning the majority of its population—are looking positively to the future and what it will bring, other places envisage lengthy and painful processes of economic and social decline. Generally, that is described as social and economic disparities and asymmetric economic growth perspectives. However, if wealth is increasingly concentrated to specific groups in society and specific types of cities and regions, the societal, economic and territorial fragmentation and the political consequences of this concentration process may undermine the fundament on which urban areas prosper.

Trends accelerating further fragmentation, therefore, threaten EU cohesion (in all its formats), develop into a fundamental threat for overall European integration and put Europe's potential to respond to globalisation at risk. Recent discussions about financial support mechanisms for losers under globalisation and the economic crises, about joint approaches to the refugee issue and about solidarity in general, point in the same direction. It seems that European integration will not be the main trend for the forthcoming years. On the contrary, the European integration process could come to a halt or might even be reversed (ESPON 2017). Thinking about possible disintegrations, one needs, however, to be aware that they would lead to reductions in economic growth in all regions (ESPON 2017, 2018).

Summing up, the increasing concentration to urban areas makes sense from an economic perspective. However, it risks amplifying societal and increasingly also territorial fragmentation processes in Europe. These fragmentations divide thriving urban areas with largely positive outlooks to the future from less well-positioned places often in rural areas or areas of economic decline with meagre hopes for a bright future. In the long run, this emerging 'geography of future perspectives' risks to threaten cohesion and European integration, and may even undermine the basis on which the thriving urban areas build their success.

9.5 In Search for Alternative Narratives

Concentration leading to societal and territorial fragmentation, which, in turn, risks the collapse of European integration and the European economic model, cannot be a narrative without alternatives. What could be done to break the line of developments and avoid increasing fragmentation or at least the negative consequences of it? Let's cast some light on three different ideas for alternative futures:

(1) The first alternative starts with the assumption that the problem needs to be solved at its very roots. That is the economic and political system which spurs concentration developments and thus leads to fragmentation. What is needed

is a revolution, at least a revolution of our economic system. There are many theories and alternative ideas around (see e.g. Burmeister et al. 2018; Lange 2018; Randers 2012; Rosa et al. 2018; Rosling et al. 2018; Welzer 2016). Let's just pick one. The FP7 project FLAGSHIP explored grand societal challenges in Europe and developed alternative visions for Europe. One of them, 'Metamorphosis', is based on a fundamental socio-ecological paradigm shift including a move to consequent circular and sharing economy (Böhme and Lüer 2016; Lüer et al. 2015). The territorial implications of this alternative vision have been analysed and show that current fragmentation trends can at least be mitigated under the assumptions of this radical paradigm shift coming close to a revolution of our economic model (Böhme and Lüer 2017). The main urban structures and territorial patterns will remain but new governance and economic approaches will allow for better policy responses. Whether this revolutionary alternative is likely or even realistic remains, however, an open question.

(2) The second alternative starts with the assumption that we may not be able to change our economic system. However, it might be possible to change mindsets to embrace diversity more wholeheartedly. Sure, already today diversity is in most parts of Europe perceived as positive. However, when it comes to personal development perspectives there is a pretty narrow corridor of possible alternatives which are generally perceived as 'positive' or 'successful'. Imagine for a moment, a Europe, where the societal perspective of a 'successful' member of society is not only centred on its continuous increase of personal economic wealth including the corresponding status symbols. Instead of a streamlined design of the personal life, changes and disruptions are welcome. Thus focusing on economic prosperity and moving to the places linked to that in some phases of working life can be mixed with phases focusing on other values and moving to places where this can be done, e.g. more quite disconnected rural areas. Today we see some seeds of such developments, e.g. when young artists or IT people set up co-working (and co-living places) in deeply rural areas. The rationales for this range from more affordable locations, via less distraction and thus more focused work on creative processes, to the search for alternative lifestyles and lifestyles which are closer to nature. Can we change our mindsets to the degree that we go back and forward between different types of lifestyles in different phases of life—depending on the needs one has in a certain phase of life—without that being perceived as 'anormal'? Following the argument of Goodhart (2017), this would mean that the majority of the population would need to transform from 'Somewheres' rooted in their communities to 'Anywheres' being more footless and flexible.

(3) The third alternative starts with the assumption that we may neither change our economic system nor our mindsets but that fragmentation can be bridged by increasing cooperation. The challenges brought about by fragmentation are also linked to the fact that today's development challenges and potential can no longer be mastered by decision makers in charge of individual territories, be it municipalities, regions or countries. The high levels of territorial interdependencies and interaction imply that for almost any development issue, territorial

impacts extend beyond administrative borders and decisions at different administrative levels and in different territorial units need to play together. (Mehlbye and Böhme 2017) Such interdependencies ranging from urban to rural, cross-border or macro-regional and transnational shape the territorial development in Europe. Overall there is more need for functional and integrated approaches. Challenges increasingly overlap and need to be addressed on different levels. Future-oriented governance arrangements and approaches need to adjust to new contexts, go beyond administrative borders and address, i.e. urban–rural linkages from an integrated perspective (Lüer and Holstein 2016).

The functional integration of urban areas can take place at different levels: global, European, cross-border and transnational, regional and metropolitan, urban as well as neighbourhood level:

- **Functional integration and cooperation at global level** refers to cities being nodes in transport, global financial systems, hotspots of research institutions and innovation or places attracting global tourism (ESPON 2013). Territorial cooperation in city networks and functional regions can be identified as 'territorial keys' illustrating which aspects of European territorial development are especially relevant in order to make policy interventions more efficient (Böhme et al. 2011).
- **Urban networking at cross-border and transnational level** would encourage partnerships and cooperation between urban regions and contribute to a more balanced development in Europe. At regional and metropolitan level, the mismatch between the real city and political delineations is increasing, while at urban level the functional integration refers to the integrated urban development. Through cooperation at that level, cross-border obstacles may be removed and regions may work on a joint objective, while economies of scale would also be increased. The Services of General Interest could overall also be improved across regions, through functional cooperation of urban areas (ESPON 2017).

In several places in the EU, cities, municipalities and regions are already cooperating with one another across administrative borders to gain more than by staying on their own. At the same time, decision makers express a growing interest in looking for opportunities beyond their territorial borders and address the function of larger areas.

As the geography of different issues vary, such networks of cooperation need to be established on different scales. Even the largest metropolises and most global cities are not globalised entirely. Only certain parts are embedded in global or international networks while other parts are integrated on neighbourhood, local, regional or national scales (Vöckler 2015). This intra- and inter-urban heterogeneity and complexity needs to be reflected by future cooperation networks. Top-down support needs to be provided to establish and maintain such bottom-up networks, e.g. by developing suitable framework conditions (funding, legislation, expertise, infrastructure, etc.). This includes the empowerment of urban areas and the creation of test beds for experimentation (Lüer and Holstein 2016).

A revolution of our economic system or a radical change of our mindsets remains rather unlikely alternatives to break the fragmentation trend. Increased territorial cooperation is the solution to avoid the most negative impacts of fragmentation. Still, functional urban cooperation is a challenge for cities as they would need to look beyond their borders. However, this way, cities can address challenges more efficiently and improve their prospects working together on joint matters. Bringing the urban concentration and the functional perspective together, it seems that the economic geography of Europe can change and find answers to becoming a European geography of promising future perspectives.

References

Bain & Company (2016) Spatial economics: the declining cost of distance. The next big economic shift will reshape industries, social patterns and the global economy. Bain & Company, New York and Dallas

Barca F, McCann P, Rodríguez-Pose A (2012) The case for regional development intervention: place-based versus place-neutral approaches. J Reg Sci 52(1):134–152

Burmeister K, Fink A, Schulz-Montag B, Steinmüller K (2018) Deutschland neu denken: acht Szenarien für unsere Zukunft. Oekom Verlag, München

Böhme K, Lüer C (2016) Europe's territorial futures between daydreams and nightmares. In: Europa XXI. Territorial uncertainty as a challenge for regional policy in Europe, vol 30. pp 5–22

Böhme K, Lüer C (2017) Assessing territorial impacts of future trends. In: Medeiros E (ed) Uncovering the territorial dimension of European Union Cohesion Policy. Routledge, Abingdon & Oxon, pp 161–183

Böhme K, Toptsidou M (2017) The future of Europe and its territories. Spatial Foresight Brief No. 2017:7. Spatial Foresight, Luxembourg

Böhme K, Doucet P, Komornicki T, Zaucha J. Świątek D (2011) How to strengthen the territorial dimension of 'Europe 2020' and EU Cohesion Policy. Report prepared at the request of the Polish EU Presidency, Polish Ministry of Regional Development, Warsaw. Retrieved from http://www.spatialforesight.eu/tl_files/files/editors/dokumente/Report_Final_110913.pdf

Böhme K, Antikainen J, Zillmer S, Hans S, Pyykkonen S (2016) Looking towards 2030: preparing the Baltic Sea Region for the future. Swedish Agency for Economic and Regional Growth (Tillväxtverket), Stockholm. Retrieved from http://www.strategyforum2016.eu/media/reports/looking-towards-2030-report-33885447

Dijkstra L, Garcilazo E, McCann P (2013) The economic performance of European cities and city regions: myths and realities. Eur Plan Stud 21(3):334–354. https://doi.org/10.1080/09654313.2012.716245

ESPON (2005) ESPON 1.1.1 potentials for polycentric development in Europe. Project report. August 2004. Revised version—March 2005, ESPON, Luxembourg

ESPON (2007) ESPON project 1.4.3 study on urban functions. Final Report. March 2007. ESPON, Luxembourg

ESPON (2013) Territorial dynamics in Europe: gateway functions in cities. ESPON, Luxembourg

ESPON (2014a) Territorial dynamics in Europe: evidence for a European urban agenda. ESPON, Luxembourg

ESPON (2014b) TOWN small and medium sized towns in their functional territorial context. Applied research 23 Jan 2013. Final Report. Version 30 Aug 2014. ESPON, Luxembourg

ESPON (2017) European territorial review. Territorial cooperation for the future of Europe, ESPON, Luxembourg. Retrieved from https://territorial-review.espon.eu/

ESPON (2018) Possible European territorial futures. ESPON, Luxembourg. Retrieved from https://www.espon.eu/territorial-futures

European Commission (ed) (2007) Leipzig charter on sustainable European cities. Retrieved from http://ec.europa.eu/regional_policy/archive/themes/urban/leipzig_charter.pdf
European Commission and UN-HABITAT (2016) The state of European Cities 2016. In: Cities leading the way to a better future. Publications Office of the European Union, Luxembourg
European Strategy and Policy Analysis System (2015) Global trends to 2030: can the EU meet the challenges ahead? European strategy and policy analysis system
Eurostat (2016) Urban Europe. Statistics on cities, towns and suburbs. Office for Official Publications of the European Union, Luxembourg
Glenn JC, Florescu E, The Millennium Project Team (2015) 2015–16 state of the future. The Millennium Project, Washington D.C
Goodhart D (2017) The road to somewhere: the populist revolt and the future of politics. Hurst & Company, London
Lange S (2018) Smarte grüne welt? Digitalisierung zwischen Überwachung, Konsum und Nachhaltigkeit. Oekom Verlag, München
Latvian Presidency of the Council of the European Union (2015) Declaration of Ministers towards the EU Urban Agenda. In: Informal meeting of EU ministers responsible for territorial cohesion and urban matters, Riga, 10 June 2015
Leunig T (2008) The regeneration game is up. Retrieved from https://www.theguardian.com/commentisfree/2008/aug/13/regeneration.conservatives
Lüer C, Holstein F (2016) Functional integration as a tool for urban areas to deliver territorial cohesion. In: EUROCITIES, making cities. Visions for an urban future. pp 74–83
Lüer C, Böhme K, Jæger S, Hans S, Madeira N, Holstein F, Toptsidou M, Tulumello S, Bina O, Ferrão J (2015) Report on territorial impact projections. Final deliverable (D7.2) of work package 7 on Territorial Governance, FLAGSHIP project, 7th Framework Programme (FP7)
MSPTD (2011) Territorial agenda of the European Union 2020. Towards an inclusive, smart and sustainable Europe of diverse regions. Agreed at the informal ministerial meeting of ministers responsible for spatial planning and territorial development on 19th May 2011 Gödöllő, Hungary
MUMEU (2016) Urban agenda for the EU. Pact of Amsterdam. Agreed at the informal meeting of EU ministers responsible for urban matters on 30 May 2016 in Amsterdam, the Netherlands. Retrieved from https://ec.europa.eu/futurium/en/system/files/ged/pact-of-amsterdam_en.pdf
Mehlbye P, Böhme K (2017) More territorial cooperation post 2020? A contribution to the debate of future EU cohesion policy, Spatial Foresight Brief No. 2017:8. Spatial Foresight, Luxembourg
National Intelligence Council (2012) Global Trends 2030: Alternative worlds
OECD (2006) Competitive cities in the global economy, Organisation for Economic Co-operation and Development, Paris. Retrieved from http://search.ebscohost.com/login.aspx?direct=true&scope=site&db=nlebk&db=nlabk&AN=184113
PBL Netherlands Environmental Assessment Agency (2016) Cities in Europe: facts and figures on cities and urban areas. PBL Publications, The Hague
Randers J (2012) 2052: a global forecast for the next forty years. Chelsea Green Pub, White River Junction, Vermont
Rodríguez-Pose A (2017) The revenge of the places that don't matter (and what to do about it), in Cambridge. J Reg Econ Soc 11:189–209
Rosa H, Kipping K, Paech N (2018) Anders wachsen! Von der Krise der kapitalistischen Wachstumsgesellschaft und Ansätzen einer Transformation (Becker M, Reinicke M, eds). oekom, München
Rosling H, Rosling O, Rönnlund AR (2018) Factfulness: ten reasons we're wrong about the world–and why things are better than you think, 1st edn. Flatiron Books, New York
Réchard D, Noonan E, Schmertzing L, Windle-Wehrle F, Frey A, Cesluk-Grajewski M (2016) Global trendometer. essays on medium- and long-term global trends. European Parliamentary Research Service, Brussels
Schwab K (2017) The fourth industrial revolution. Penguin Books
Vöckler K (2015) Die kommende "Welt-Polis". Inf zur Raumentwickl 4:323–333

Welzer H (2016) Die smarte Diktatur: der Angriff auf unsere Freiheit, 3rd edn. S. Fischer, Frankfurt am Main

World Bank (2018) *Rethinking lagging regions. Using cohesion policy to deliver on the potential of Europe's regions*. World Bank Report on the European Union

Kai Böhme is Founder and Director of Spatial Foresight, a private consultancy and independent think tank in the area of European territorial policies and research, with team members located in Luxembourg, Germany, France and Spain. Kai holds Ph.D. in Management Science from the University of Nijmegen (Netherlands) and a Master's in Spatial Planning from the University of Dortmund (Germany). He specialises in European regional and territorial research and policies, international comparative studies in the fields of regional development policies, spatial planning, and in the territorial impacts of sector policies. He has a truly European background and considerable experience in policy advice at the European and national level as well as in the management of international applied research and consultancy projects.

Christian Lüer is a Senior Consultant at Spatial Foresight and a Visiting Lecturer in the Department of Geography at the University of Münster, Germany. He has a particular interest in territorial governance, spatial planning, cross-border and transnational cooperation, and territorial impact analysis. Over the past years, he has been involved in research and consultancy projects for regional, national, cross-border, transnational and EU organisations.

Maria Toptsidou is a Junior Consultant at Spatial Foresight. Among her main professional interests lie the European territorial cooperation, territorial governance and European Cohesion Policy. At Spatial Foresight, she has been involved in projects for DG Regio, the European Parliament and the European Investment Bank. She contributed to numerous ESPON publications, Interreg programmes evaluations and supported the preparation of the EUSAIR and EUSALP macro-regional strategies.

Index

E. Medeiros (ed.), *Territorial Cohesion*, The Urban Book Series,
https://doi.org/10.1007/978-3-030-03386-6

Printed by Printforce, the Netherlands